MICHIGAN
ORCHIDS

An Illustrated Guide to the Wild Orchids of Michigan

ARETHUSA BULBOSA

STEVE CHADDE

and

MARJORIE T. BINGHAM

MICHIGAN ORCHIDS
An Illustrated Guide to the Wild Orchids of Michigan

Steve Chadde and Marjorie T. Bingham

A Pathfinder Field Guide, published by Orchard Innovations

ISBN 978-1951682705

The author can be reached at: steve@orchardinnovations.com

VER. 1 (9/1/2022)

CONTENTS

PREFACE

MICHIGAN ORCHIDS is written for the lay naturalist who wishes to know more about the state's rare and beautiful native orchids (and one introduced orchid, *Epipactis helleborine*). The first edition of this book was published by Marjorie Bingham in 1939 as *Orchids of Michigan*. The current work has been completely revised and updated to reflect our current understanding of orchid taxonomy and distribution within Michigan. The descriptions of orchids, while specific, are not technical in nature. I have attempted to keep much of the flavor of Marjorie Bingham's original work, while providing additional information to make identification of the state's orchids as easy and as accurate as possible.

The species are arranged in alphabetical order first by genus, then by their scientific, or botanical, name. These names largely follow those of the Integrated Taxonomic Information System (*itis.gov*), and are intended to reflect the latest taxonomic research into genera and species. In some cases (especially in the Ladies'-Tresses, *Spiranthes*), recent studies have resulted in a number of changed names and newly defined species, and those are incorporated here. Common names, while not standardized, are those that seemed to be most popularly or widely used in our region at the time of publication.

—STEVE CHADDE

INTRODUCTION

ORCHIDS ARE WILD FLOWER ARISTOCRATS. They are found in moist and not too cool climates throughout the world. Rather particular in their requirements of soil and moisture, orchids, though widely distributed, are found only in the special places that definitely meet their needs. Occasionally certain species may be plentiful in a restricted area, but none are generally abundant. When first published in 1939, *Orchids of Michigan* provided an estimate of 10,000 orchid species worldwide. Today, that number has grown to about 28,000 species in more than 700 genera, and although cosmopolitan in their range, they are most abundant in the tropics. In the United States and Canada, 350 species of orchid are known, and Michigan is home to 19 genera, 58 species, and six hybrids.

Orchids are all perennial herbs, many of which grow from bulbous enlargements at the base of the stem. These enlargements suggested the name, orchid, from the Greek, *orchis*, or testicle, because of the shape of the root tubers in some species of the genus *Orchis*. While a large proportion of orchids are terrestrial, a great number of tropical orchids grow on trees. They are not parasitic, however, for having green leaves they make their own food and merely use trees as a support upon which to grow. Epiphytic or tree-inhabiting orchids absorb moisture through their aerial roots, which have a specialized outer spongy layer of empty cells kept open by spiral thickenings in the walls. Moisture is transmitted to the inner tissues through this spongy layer. Most tropical orchids are epiphytes.

Terrestrial orchids, while having all the organs common to the higher plants, are also dependent on symbiotic relationships between their roots and a variety of fungi. Orchid mycorrhizae, the fungi that have a symbiotic relationship with the orchid, are especially important during orchid germination, as an orchid seed has virtually no energy reserve and obtains its nutrients from the fungus. The coralroots, however, have neither green leaves nor true roots but an underground fungus-infested rhizome. Since they live on decayed organic matter, they are considered saprophytes.

Orchids possessed of a tuberous rhizome have slender roots growing from it, while those with slender rhizomes may be supplied with tuberous roots, as noted by Holm. Often there is no difference in the diameter of roots and rhizomes. Slender rhizomes may terminate in slender roots. Holm states that the development of varied roots and rhizomes does not seem to be dependent on the nature of the environment, since species with tuberous rhizomes may occur in open bogs as well as in deep shaded woods. Species with slender rhizomes may be found in bogs, ravines, dry fields, clearings, or thickets.

Equipped with fibrous, fleshy or bulbous roots, herbaceous stems, parallel-veined leaves, and the parts of their flowers in threes, orchids resemble their close relatives, the members of the lily family (and its modern-day-divisions). The floral structure, however, and particularly the organs of reproduction, show the most highly specialized adaptations for cross pollination and fertilization of the flowering plants. The union of stamens and pistil into one organ, the column, distinguishes this family from all related ones.

The progenitors of our present orchids were probably terrestrial natives of the great tropical forests. Like the orchids of today, they were insect pollinated and produced numerous minute primitive seeds but did not have their stamens and pistil united into a column, in the opinion of Rolfe.

A few orchids are useful to humans in special ways. The fruit of a tropical climbing orchid, *Vanilla planifolia,* is the source of our vanilla flavoring. From the roots of European species of *Orchis* and from East Indian species of *Eulophia,* a ground flour, *salep,* is obtained. Since salep is also rich in a mucilaginous substance, *bassorin,* it has long been used medicinally like gum-arabic as a demulcent.

A few other orchids have been utilized medicinally. An infusion of the powdered root of the yellow lady's-slippers, *Cypripedium parviflorum,* collected after flowering, was an official drug used as a nerve medicine. According to Alice Henkel in *Wild Medicinal Plants of the United States,* the rhizome of one of the coralroots, *Corallorhiza odontorhiza,* and the entire plant of two of the rattlesnake plantains, *Goodyera pubescens* and *G. repens,* were used as unofficial drugs. Native Americans made use of orchid roots and leaves to treat many ailments.

Some of the lady's-slipper orchids are irritating to the skin if touched. Early on, MacDougal in 1894 and Chestnut in 1898 pointed out that the hairs covering the stem and leaves of the Showy Lady's Slipper and of the Yellow Lady's-Slippers, were the cause of skin irritation similar to that caused by poison ivy and poison sumac. The hairs are invested with a filamentous fungus and although the exact cause of poisoning is not known, it is believed to be due either to the presence of this fungus, to surface irritation caused by the contents of the glandular tip of the hair, or to the action of the acid contents of the hair when it pierces the skin. The irritating action increases with the seasonal development of the plant and reaches its maximum with the formation of the seed pod. MacDougal suggests that this is an efficient device for protecting the reproductive organs during the period from pollination to the maturity of the seeds. No specific counter-irritant for 'Cypripedium poisoning' has been developed.

PARTS OF ORCHIDS AND THEIR FUNCTIONS

It is by the nature of their flowers that orchids are best known, although they are distinguished by root and leaf characters already discussed. The sepals are generally not green but are colored like the petals. Two of the sepals are alike, but the third is larger and forms a hood over the petals. Similarly, one of the three petals is larger than the other two and is known as the lip (or labellum). It may be expanded and fringed, forked, or pouched, and is often brilliantly colored. It sometimes extends backward into a spur which, in some species, is long and slender; in others it is quite short.

Within the floral envelope the stamens and style are united to form a thickened organ known as the *column;* the structure of the column varies. There may be from one to three anthers. Often there are but two fertile ones, the third forming a sterile leaf-like structure which hangs over the stigma. Sometimes there is only one fertile anther with two pollen masses, one on either side of the stigma. The column itself may be elongated and project over the stigma.

Orchid pollen differs from the fine, granular, loose, powdery grains common to other plants. it is instead in waxy cohesive masses, its grains connected with elastic cobwebby threads which may permit the entire mass to be stretched to four or five times its length and to recover its original shape when released. These pollen masses are known as *pollinia* and are carried away as a whole rather than as separate grains.

Orchids have their three pistils united; the upper and anterior surfaces of two of them form the two stigmas, which are often so completely joined that they appear as one. The upper stigma is so modified that in many orchids it does not even resemble a stigma. It is called the *rostellum* and is a specialized organ which, when mature, either includes or is altogether formed of viscid matter. In many species the pollen masses are firmly attached to a portion of the exterior membrane, so that it is removed with them when insects visit the flowers.

Orchids are insect pollinated, and their floral structures are especially adapted to the visits of certain insects. So specialized are they that, failing to attract the right visitors by their nectar, color, and fringes, they produce no seed, for orchids, in general, have no provision for self-pollination.

The lip of the flower offers a landing place for insects, which are attracted not only by color but also by fragrance. There is a wide difference of opinion regarding the odor of orchids, and the scent varies within each genus. Some species, such as Prairie Ladies'-Tresses, are definitely strongly scented, while other *Spiranthes,* such as Sphinx Ladies'-Tresses, have little to none.

Orchid flowers, more variously shaped than those of any other plant family, have many devices for achieving their ends. They must by some means attract the specific insects that can pollinate them. So necessary are these insect visits that orchid flowers remain fresh an unusually long time and are noted for their longevity. Some kinds keep fresh for a month or longer when they are not pollinated. Once pollination has occurred, however, the flower wilts.

The form and structure of orchid blooms is correlated and adapted in minutest detail to the form and habits of desired insect visitors. Darwin states, "It is a safe generalization that species with a short and not very narrow nectary are fertilized by bees and flies while those with a much elongated nectary, or one having a very narrow entrance are fertilized by butterflies or moths, these having long and thin proboscides." The length of the nectary is undoubtedly adapted to the length of tongue of the visiting insect. A tropical species secreting nectar at the bottom of a nectary eleven inches long is pollinated by a sphinx moth with an eleven-inch tongue. Flowers with short nectaries are pollinated by insects having short tongues. This coordination is necessary because the gathering of nectar is coincident with pollination. If the anatomy of the insect and the structure of the flower do not coincide, fertilization cannot result.

In some orchids the nectar is free, that is, it lies openly within the nectary. In other species insects must bore through the inner membrane of the nectaries to obtain it. This difference in secretion of nectar is correlated with the particular mechanism of each flower to assure pollination. The cement of the viscid discs by which pollen masses adhere to the bodies of visiting insects is kept moist by a liquid as long as the discs are in place, but when the discs are removed and exposed to the air the cement requires varying lengths of time to harden. Orchids whose nectar is easily and quickly available have cement on their discs which hardens rapidly. Pollen masses are, therefore, speedily and firmly fixed in correct position on the body of the visiting insect. This is im-

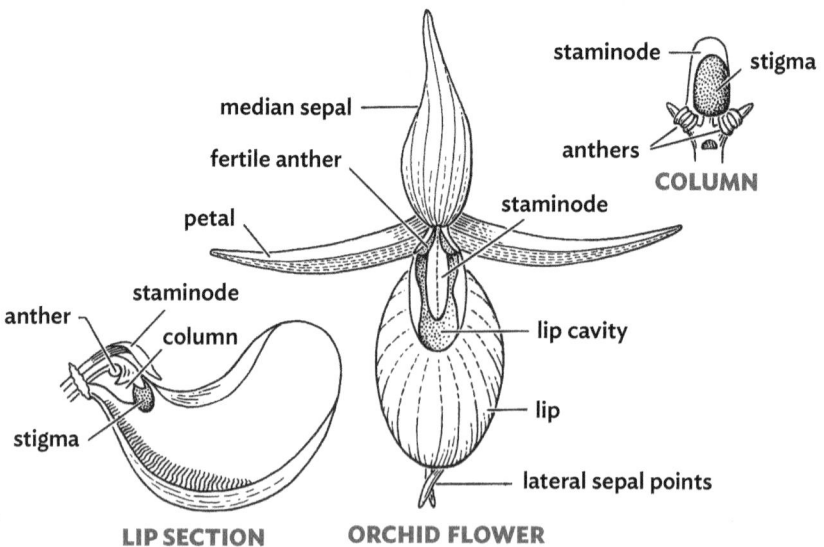

LIP SECTION   ORCHID FLOWER

portant, for if they slipped on one side or the other they would not come into contact with the right portion of the pistil of the flower to which the insect paid its next visit. In orchids where insects must bore holes at several points through inner membranes of the nectary to obtain sufficient nectar, the cement of the adhesive discs hardens slowly. In these species it is necessary that the insect be delayed long enough for the cement to harden before he leaves the flower. Otherwise the pollen masses would be lost. Particular mechanisms of the Michigan orchids are noted in the accounts of the species.

The seeds maturing as the result of fertilization are very numerous and exceedingly minute. They have loose membraneous coats, no endosperm or stored food, and in the embryo itself one often cannot discern any trace of external organs. The seeds are borne around the sides of a pod or capsule, which splits open by means of valves.

ORCHID CONSERVATION

It is strangely true that in their zeal humans often destroy that for which they cares most. Nowhere is the illustration of this truism more obvious than in the realm of nature. Intensely appreciative of the beauty of wild flowers, many of their admirers instinctively pick them on sight realizing too late, if at all, that in the act of picking they have contributed to the ultimate destruction of the species from a particular place.

Since the leaves of plants are the organs in which their food is manufactured that, when transformed into energy, makes possible their continued growth, it is imperative that the activities of these constructive organs must not be impaired. Food not needed to maintain the season's growth is transmitted to the plant's underground storage organs: bulbs, tubers, or fleshy roots, where it is held in reserve to start growth the following season. When the leaves of wild flowers are picked, the plant's chance to make and store food for the next season's growth is destroyed. Because many wild flowers have leafy stems, their blossoms cannot be picked without the accompanying leaves.

Native orchids are all perennial herbs and are dependent on stored food within their underground reservoirs to perpetuate their species. Since many of them depend not only on enlargement and multiplication of their bulbs and tubers but also on the production of seed to continue their race, it is essential that blossoms be permitted to remain and mature on their stems so that seeds may result from fertilization. Picking flowers prevents the formation of seed. While many wild flowers are so plentiful that one may occasionally disregard these facts, the preservation of native orchids permits no exceptions.

While many species, according to Denslow, (1927), are "tolerant for years of changed conditions, they cannot survive ultimately any radical alterations of their environment." Thoughtless picking speeds destruction. Likewise, transplanting from an orchid's native habitat to a backyard flower garden should never be attempted, as the chance of success is minimal. There are a number of reputable nurseries who propagate some of our native orchids, and with proper site preparation and care, it is possible to grow them in the home landscape.

Today, a number of orchids are protected under the Endangered Species Act of the State of Michigan, signed into law in 1994. Currently there are five Michigan endangered orchids, six threatened orchids, and three species of 'special concern,' which, while not legally protected, are monitored as their numbers are apparently declining in the state (species with special status are noted in the text). One species, the Prairie White Fringed Orchid (*Platanthera leucophaea*), is also listed as threatened at the federal level due to its rangewide rarity. In addition, as a result of heavy collecting, Michigan has laws protecting various groups of wildflowers, including the state's native orchids. More information is available from the Michigan Natural Features Inventory (*mnfi.anr.msu.edu/species/plants*)

Orchids have long fascinated the imagination with their exotic beauty. Shown above is a colored engraving of Loesel's Twayblade (*Liparis loeselii*), from an early book, *Flora Europaea inchoata,* published in 1797.

PUTTYROOT
Aplectrum

AS EARLY SETTLERS found they were able to make a sticky paste which would mend broken pottery by grinding the roots of this orchid and mixing the powder they obtained with water, it became known at **Puttyroot**. Of similar homely origin is the widely used name **Adam and Eve**. This common name is derived from the plant's paired bulbs and originated with the southern African-Americans who used the bulbs as charms and attempted to divine the future with them. The scientific name *Aplectrum*, meaning "spurless," comes from the Greek and obviously refers to the absence of that organ.

PUTTYROOT
Aplectrum hyemale (Muhl. ex Willd.) Torr.

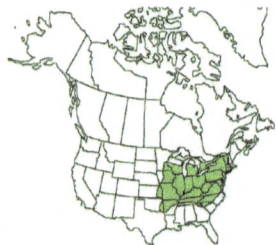

Puttyroot takes advantage of the limited light available in its deciduous forest habitat by having its leaf present from the fall, through winter, and into early spring, when overhead trees are leafless. As flowering time nears, the leaves wither away.

Puttyroot, like Calypso, begins its life cycle in late summer, when it sends up from its biennial tuber a wrinkled, plaited, bluish-green, oval leaf, which lasts through the winter and vanishes in late spring.

The slender flowering stalk, nine to twelve inches in height, bears about ten flowers, each one a ½ inch long, in a loose raceme from three to four inches in length. Inconspicuously colored greenish-madder or sometimes yellow, the fragrant flowers bloom in May and June.

The purplish-green petals and sepals are narrow, tipped with dull purple. While the sepals are spreading and free, the petals project side by side over the column and lip, which is white with purple markings. It is three lobed, and although the two lateral ones are curled upward, the middle one is somewhat expanded and has wavy margins. A flattened ridge, bordered on each side by a low crest, marks the median line of the lip.

The bulbs from which Puttyroot grows are not true bulbs but corms like those of Crocus. They are fleshy and round, often an inch in diameter, and live for several years. They send out offsets which are attached to the parent corm by fleshy ligaments. Because they persist for two years or more, when uncovered two or three of them are seen to be fastened together. This habit gave the orchid its name 'Adam and Eve'.

Seeming to choose moderately rich but neutral soil of deciduous forests, Puttyroot shows a preference for the shade of beech and maple. Like the rattlesnake plantains, it is not free flowering, and one may find few blossoms among a colony of many wrinkled, solitary leaves.

PUTTYROOT
Aplectrum hyemale
ALLEFANT

young flower stem
VA STATE PARKS

PUTTYROOT
Aplectrum hyemale
CHOESS

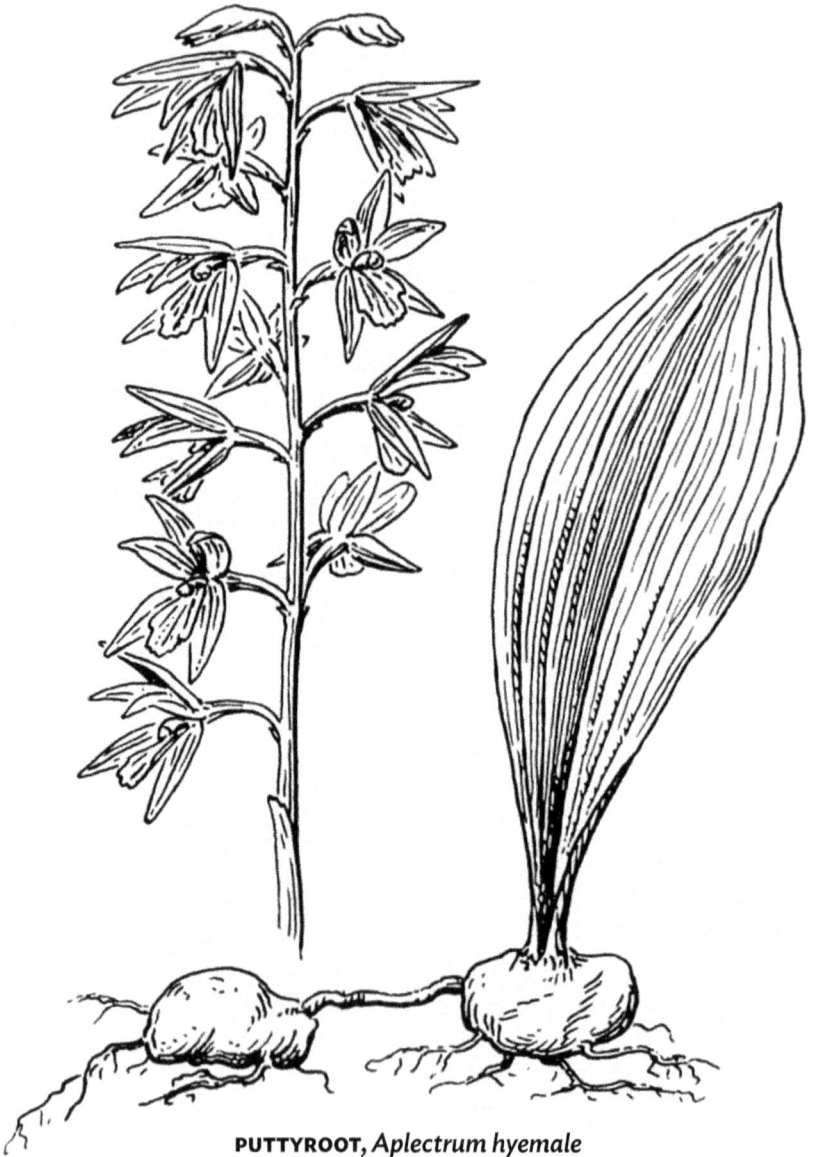

PUTTYROOT, *Aplectrum hyemale*

ARETHUSA
Arethusa

THERE IS BUT A SINGLE SPECIES of *Arethusa* throughout the world, and found only in North America. However, *Arethusa bulbosa* is closely related to *Eleorchis japonica* of Japan, which historically was treated as *Arethusa japonica*.

ARETHUSA
Arethusa bulbosa L.

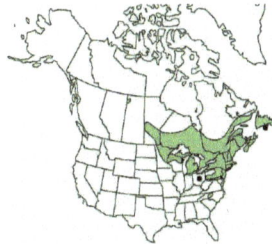

Flower of the gods, **Arethusa** was named for a wood nymph of Greek mythology. Through June and July on sphagnum hummocks of open acid bogs, this incomparable flower of glistening crystalline texture may be found throughout Michigan. As far north as Isle Royale and as far south as Oakland County this superb orchid thrives. It cannot be said to grow in great abundance, however, and like all orchids, when found it should not be picked.

Rising from deep sphagnum hummocks among a tangle of cranberry vines, Arethusa's single magenta flower is borne at the tip of its slender stem, which rises five to twelve inches from a small firm white bulb. Although leafless when in bloom, a single, grass-like, sheathing leaf appears subsequent to the blooming season and often grows to a length of six inches.

The flower unfolds between a pair of small scales. As in many orchids, the petals and the upper sepal form an arching hood extending over the column, while the lower sepals, lance-shaped, curve backward and upward. The broad tongue-like lip also curves upward but only to roll out and down, displaying its pinkish-white, wavy-margined, crested lip. Three rows of short, white, fleshy, yellow and purple-tinged hairs adorn the lip, whose margin is rose-purple and whose floor is spotted and streaked with madder.

Arethusa flowers slightly earlier than *Calopogon tuberosus* and *Pogonia ophioglossoides*, which often grow together in the same boggy habitats.

To prevent self-pollination and to insure the cross fertilization of its flowers, Arethusa has adopted some of the same methods used by Rose Pogonia. The column, spread out like a petal under the over-arching hood of petals and upper sepal, covers the anther, which, in turn, lies immediately over and projects beyond the stigma.

The tightly closed anther consists of two cells, each holding two pollen masses. As the insect advances along the crested lip to the shallow nectary lying at its base, it brushes against the sticky stigma, and, having to retreat from the nectar by backing out of the orchid's throat, it touches in its egress the tip of the anther. With a spring it opens, depositing pollen masses on its head or thorax.

Little would one expect to discover this orchid in saline soil; nevertheless it is so abundant on some parts of the northern New England coast that in June the salt marshes are purple with them.

DAVID MCCORQUODALE

ARETHUSA
Arethusa bulbosa

JOSHUA MAYER

ARETHUSA
Arethusa bulbosa

JOSHUA MAYER

GRASS PINK
Calopogon

FIVE SPECIES OF GRASS PINK are native to the eastern part of the United States, Cuba, and the Bahamas, but only one occurs in Michigan. Our species is the commonest of the three showy pink "bog" orchids: *Arethusa, Calopogon,* and *Pogonia,* especially in southern Michigan.

GRASS PINK
Calopogon tuberosus (L.) Britton, Sterns & Poggenb.

SYNONYM *Calopogon pulchellus* (Salisb.) R. Br.

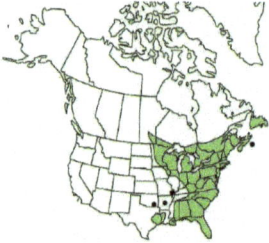

Grass Pink has the widest range of any of the five species in this North American genus: it is found across much of eastern Canada, the eastern United States, and south to the northern Caribbean.

Companion of Rose Pogonia in open wet grassy meadows, sphagnum bogs, and openings in conifer swamps, **Grass Pink** blooms during July and August. It has been found in acid soil throughout Michigan, where it appears quite impartially in both the deciduous and lake forest zones.

The most unusual feature of this flower is the inverted position of the lip. Instead of growing outward or downward from the center of the flower, the lip is the uppermost organ of all, and the sepal, which in other orchids is above and generally arching over the column, has taken the position usually occupied by the lip. The lower sepals are found where other species have their petals, and the petals have taken the place normally occupied by the sepals. The column curves outward above the middle sepal.

A beautiful flower, its floral organs, including the column, are rose-pink or magenta and have a fragile, glistening quality. It is the lip, however, which is most arresting. Narrowed and hinged at the base, its outer end is bearded with a brush of three or four parallel lines of white fleshy hairs, with magenta and yellow tips. This orchid well deserves its scientific name which, derived from the Greek, means "beautiful beard."

Nor is the superb beauty of the Grass Pink's lip its only recommendation. It is the lure which draws a series of small bees to effect pollination. As soon as an insect alights, the lip swings forward on its hinges and deposits its guest in an inverted position upon the column. Here, as described by Watts: "The back of his abdomen receives an application of sticky stigmatic fluid. As he glides past the end of the column his weight ruptures the tiny pouches which contain the pollen masses. These become attached by means of fine gossamer-like threads to the sticky spot on his back and the insect is burdened with a load which he cannot use and cannot discard until he again allows himself to be lured by false promises." While an efficient method of pollen transfer, Grass Pink gives the insect neither nectar nor edible pollen, so that it deserves the epithet given it by Watts: "a clever deceiver."

Three to twelve flowers grow in a loose zig-zagged raceme which is open enough to permit each flower ample room to expand and be individually alluring to insects. The flowers have no spurs.

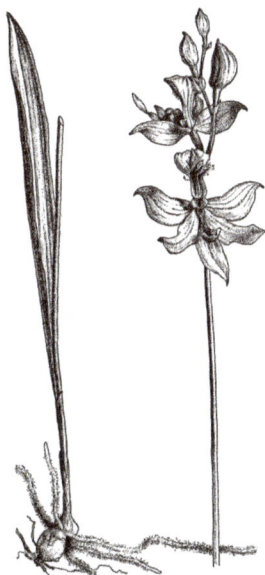

GRASS PINK
Calopogon tuberosus

A single narrow grass-like leaf sheaths the slender stem which springs from a small white bulb supplied with long fleshy roots. Each year a new bulb is produced from which the ensuing season's growth will come. So while the old bulb is discarded each year, an unbroken line is assured.

GRASS PINK
Calopogon tuberosus

ALLEFANT

CALYPSO
Calypso

NAMED FOR THE NYMPH which Ulysses found so captivating that for seven years he forgot his Grecian home, Calypso is deemed by many who appreciate wild orchids as the most beautiful of them all. It was a great favorite of the naturalist John Muir, who was so overwhelmed by its beauty that he wrote: "How long I sat beside Calypso I don't know. Hunger and weariness vanished and only after the sun was low in the west I splashed through the swamp, exhilarated as if never more to feel mortal care." In North America, Calypso is a lone representative of its group, but it has relatives in Russia, Lapland, and Northern Europe, as well as in the East Indies.

CALYPSO
Calypso bulbosa (L.) Oakes

STATUS Michigan Threatened

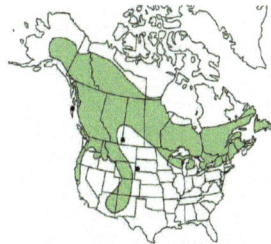

Calypso, while forming large colonies on Isle Royale, in the Lower Peninsula this species is much rarer, especially away from the shores of the Great Lakes.

Often called 'Fairy-Slipper' and 'Hider of the North', this boreal flower was originally classed as a lady's-slipper by Linnaeus. It somewhat resembles that group of pouched orchids, and by some it is still provincially called a lady's-slipper. The fancied similarity is but superficial, however, and Calypso stands apart.

This enchanting orchid, so surpassingly exquisite that it beggars description, dwells in primeval forest under pine and hemlock where, often in the semi-twilight, no other plant is in evidence on the forest floor. In other places it may be on a moss-covered rotting log in company with Twin-flower. Sometimes, in the thick humus of birch, spruce, and aspen cover, it competes for one's attention with the similarly colored fringed polygala. Single plants are more usual than colonies, although occasionally they occur in clumps. Cold mossy cedar swamps also claim Calypso, but in these locations it always occurs on some higher bit of ground—the base of a tree or on some old stump or log. It is rarely in pure soil but generally in leaf mold or beds of moss.

Calypso is about six inches in height, and bears a single flower at the summit of its pale purple stem.

The blossom is delicate rose-lavender, white and gold. The lance-shaped rose-colored petals and sepals are alike in spreading and ascending wing-like over the shoe-shaped translucent lip, within which are numerous madder-purple lines. The superfluous edges of the lip instead of being turned inward like those of the lady's-slippers is rolled outward in front like a glistening white apron, with two horns like two toes protruding below in mid front. The throat of the flower is marked with streaks and spots of pink, violet and cinnamon-brown. The upper front edge of the orifice is bearded with three double rows of golden yellow hair.

A single plaited basal leaf, oval in form, appears in late summer, persists through the winter, and withers soon after the orchid has bloomed. Following a brief period of apparent dormancy, a new leaf is developed. Calypso grows from a small firm white bulb which produces off-sets to assure continued propagation.

Calypso blooms through May and early June. In Michigan it occurs, rarely, in the northern part of the Lower Peninsula, in evergreen forests of the Upper Peninsula, and in the spruce-fir woods of Isle Royale.

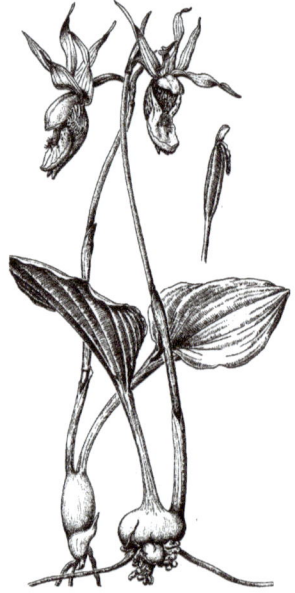

CALYPSO
Calypso bulbosa

BREWBOOKS

CALYPSO
Calypso bulbosa

CORALROOT
Corallorhiza

THE CORALROOTS ARE THE ONLY WILD ORCHIDS native to Michigan which have no green coloring and consequently do not make their own food. They are true saprophytes, living on decaying organic matter, and are sometimes classified by orchid enthusiasts as "poor relations." Although they received their name from the coral-like formation of their roots, they really have no roots but a thick much-branched underground stem. Their leaves are reduced to sheaths and are marked with the brownish, yellow or purple tones characteristic of the species. The flowers grow somewhat stiffly in loose racemes and, while unattractive en masse, are individually, when examined under magnification, amazingly beautiful.

Coralroots are 'myco-heterotrophic,' that is, they obtain nutrients for growth not from photo-synthesis, but from decaying organic matter by using mycorrhizal fungi.

In all the coralroots the petals and sepals are of about the same length, and while portions of the lower sepals are somewhat fused with the lip to form a sac-like spur, this organ is only visible in one species. The column bears an anther at its end whose tightly-fitting lid covers two paired pollen masses. On the under surface of the column lies the stigma. As an insect enters a young flower it trips open the anther lid, immediately releasing the waxy pollen masses which adhere to its head without the aid of viscid discs. Being relatively immature, the throat of the flower is not wide enough for the visiting insect to obtain nectar. However in the first mature and more widely opened flower which it visits, the pollen masses are transferred to the stigma which must be passed en route to the nectary.

About seven species of coralroot are found throughout North America. Four species and varieties of them grow in Michigan. The characters of the lip distinguish the species.

Spring coralroot, *Corallorhiza wisteriana* Conrad, is reported from northern Indiana and may one day be found in southwestern Michigan. It resembles *C. odontorhiza*, but blooms in the spring, lacks a bulbous base to the stem, and has a larger lip.

SPRING CORALROOT
Corallorhiza wisteriana
ERIC HUNT

LARGE CORALROOT
Corallorhiza maculata (Raf.) Raf.

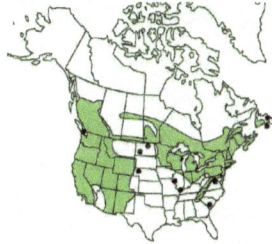

The **Large Coralroot**, whose specific name, *maculata*, refers to its spotted character, blooms from the middle of June to the middle of August, in dry well-drained upland woods of all kinds. It has a marked preference for steep slopes, and chooses, in the more boreal portions of its range, the cover of pine, spruce, fir and hemlock. Southward in the state, it is found in mixed hardwood forests.

This coralroot is taller, stouter, has larger flowers and more of them than the other species in the group. In addition to its greater size, it has several well defined marks of distinction. So spotted and lined is it with crystalline purples, browns, pinks and yellows that it is unique and of striking beauty. It is also often called **Spotted Coralroot**.

The entire plant throughout the eight to sixteen inches of its height is colored in rich shades and tones. The loosely racemed three to seven inch spike may contain nearly forty flowers but twenty is more typical.

The large blossoms are three-quarters of an inch long, and while the upper sepal and petals unite to form a hood over the column, in true orchid fashion, the lateral sepals are spreading. The white three-lobed lip is so cleft that it resembles the head of a spear. The middle lobe is squarish and blunt-tipped. Two parallel ridges follow the median line of the lip and probably function as guides to the nectary. The Large Coralroot is the only one of its group with a conspicuous spur.

When in full bloom the flowers are stiffly erect, but after they have been fertilized and the seeds begin to ripen, the pods droop.

In the past, three varieties of this species were described for plants found in Michigan's Upper Peninsula. One, var. *flavida,* differed from the typical form in color, having an absence of the madder, brown, and pink tones characteristic of the species. In this form, the stem with its sheaths is pale yellow, the flowers are lemon-yellow, and the lip is unspotted white.

An intermediate variety, var. *intermedia,* was reported for Keweenaw County. Its flowers were yellowish-brown, spotted with purple.

Another little known variety, var. *punicea,* has "no trace of brown whatever, in any part of the plant," according to Bartlett, who further stated that its stem and developing fruit were purple and that the sheathing leaves are a much paler purple than the stem. However, these varieties are today considered merely variations in forms and not true botanical varieties.

LARGE CORALROOT
Corallorhiza maculata

THOMAS GOLDBERG

SMALL CORALROOT
Corallorhiza odontorhiza (Willd.) Nutt.

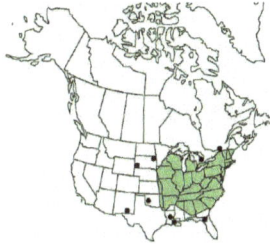

One may find this low growing orchid in open oak woods, pine plantations, and upland forests of mixed hardwoods, especially in deep humus, from August through October.

Often called the Autumn Coralroot because of the lateness of its bloom, this little noticed plant is barely seven inches tall. It has, throughout, the brownish-purple tones of the genus, but is differentiated from our other species by its greenish petals and the roundness of its white lip, which is lightly spotted and edged with violet. It is the only coralroot with a thickened bulbous base.

Small Coralroot is our only *Corallorhiza* whose stem is swollen or bulb-like at its base.

LIV MW

SMALL CORALROOT
Corallorhiza odontorhiza   JUDY GALLAGHER

STRIPED CORALROOT
Corallorhiza striata Lindl.

Plants of **Striped Coralroot** may occur as a single stalk or in large clumps of a dozen or more stems.

This is the most conspicuous and the most beautiful of all the coralroots. In neutral soil, growing at its best in limestone areas, this madder-striped orchid blooms from late May through July.

Its six- to eighteen-inch magenta stalks are sheathed with whitish, yellowish or pale purple scales and are crowned at their summit with a loose six- to eight-inch raceme of glistening madder-purple and bronze flowers.

The drooping lance-shaped petals and sepals, which may be somewhat oval or oblong, form a loose hood over the column. They are creamy yellow on the outside but crystalline white within. Three long purple stripes mark each sepal, while the petals have in addition to the three full length stripes two shorter ones. The ruby lip shades lighter toward the center of the flower and is striped like the petals and sepals. It is tongue-shaped and has a deeply grooved ridge along its median line.

To be fully appreciated, the Striped Coralroot is best seen in sunlight, when its bronze-madder is transformed into the blood-red of rubies.

STRIPED CORALROOT
Corallorhiza striata

WALTER SIEGMUND

STRIPED CORALROOT
Corallorhiza striata

JASON HOLLINGER

EARLY CORALROOT
Corallorhiza trifida Chatelain

The **Early Coralroot** is unmistakably identified by its low stature, its greenish-yellow flowers, and the square or flat tip of its lip. Growing singly or in clusters of three to eight or more plants, it thrives in deep rich forests of evergreens or hardwoods (often in deep shade), in willow and alder swamps, and in tamarack-canopied bogs. The chemistry of the soil is apparently inconsequential, for it grows as well in acid humus as in neutral or alkaline soil.

It is generally conceded that while the petals and sepals are greenish-yellow and the lip white, in some localities these basal colors may be spotted with purple. Early coralroot varies in height; sometimes it is twelve inches high, at other times and in different situations it is but six. It is not rare, but it is often overlooked.

It enjoys a moderately long blooming season since it is found as early as mid-May and as late, in the North, as August. Drooping cylindrical green pods replace the flowers and seem to prove the exception to the rule that coralroots lack green pigment. Its transcontinental range in North America is supplemented by its occurrence in Asia and Europe.

EARLY CORALROOT
Corallorhiza trifida
JOHNDAL

JOHNDAL

EARLY CORALROOT
Corallorhiza trifida

JOSHUA MAYER

LADY'S-SLIPPER
Cypripedium

TO THIS GROUP BELONG MOST OF THE FLOWERS recognized by the layperson as wild orchids. Variously known as moccasin flowers, whip-poorwill shoes, squirrel shoes, and lady's-slippers, the generic name *Cypripedium* comes from the Greek, and means, literally, "Venus' slipper."

In Michigan five species of lady's-slippers are found. While some have a marked preference for certain latitudes, others are almost impartially broad in their range. The **Pink Lady's-Slipper** is most plentiful and cosmopolitan, being found on evergreen heaths of the Keweenaw peninsula as well as in bogs of the southern counties of the state.

Darwin claimed that lady's-slippers differ from all other orchids far more than any other two genera do from each other. The inflated sac or pouch is the most conspicuous characteristic of lady's-slippers. It is variously colored in the different species and may be white, yellow, or pink. All lady's-slippers but one, the Pink, *Cypripedium acaule,* have leafy stems.

Our **Lady's-slippers** have gland-tipped, irritating hairs on the stems leaves that can cause a skin rash similar to that caused by poison ivy if plants are touched.

The two to several conspicuous leaves which sheath the plant axis are arranged alternately, have parallel veins which are frequently prominent, and are somewhat folded in appearance. The flowers are generally solitary and terminal, but when two or three are present they grow in a simple raceme. The petals and sepals are similar in texture but usually differ from each other in color. The two paired petals are free. They are generally extended and are narrower or broader than the sepals. The third petal forms an inflated sac-like lip with its edges folded inward, making an opening by which pollinating insects enter.

Bees pollinate lady's-slippers, yet there is no nectar within the lip. However, the tips of the hairs which line its inner surface secrete minute drops of slightly sticky fluid. Although bees enter by the large opening on the upper surface of the pouch they are prevented from returning the same way by the inturned edges of the lip, which are often smooth and polished. To make their exit they must crawl out by one of two other and smaller orifices. To leave the flower the insect must first brush past the stigma and then one of the anthers, so that it cannot leave without depositing on the stigma pollen with which it has entered, and subsequently taking pollen from the anther to the next flower visited. The kinds of bees which pollinate lady's-slippers do not seem to be very plentiful, for one may watch for hours and not see a single bee enter. Often other insects go in which are too large to escape through the narrow exits, so that they are trapped and held prisoner until the flower wilts or until they die.

The rarity of orchids has sometimes been attributed to scarcity of the pollinating insects which they specifically require. It is also thought that during the unusually long period, often from one to four weeks, intervening between pollination and fertilization, many fatalities may occur.

The stigma of this group of orchids is really three united into one. It has no overhanging beak, or rostellum, such as many others possess. There are two fertile anthers whose grains of pollen are not united by threes or fours, as in some other groups, neither are they tied together by elastic threads nor cemented into waxy masses. The pollen is loose and pulpy or powdery, but is coated by glutinous, sticky fluid which enables it to adhere to the dry convex stigma. In all other orchids, except Vanilla, the pollen is more or less dry and is assured of anchorage by viscous fluid excreted by the concave stigmas and the beak which overhangs them.

Lady's-Slipper seed pods are long three-angled capsules containing a quantity of infinitesimal seeds. Their roots are fibrous and come from a fleshy creeping rootstock or rhizome which has a pungent, spicy odor.

PINK LADY'S-SLIPPER
Cypripedium acaule Ait.

Differing from all the other lady's-slippers in having two broad basal leaves and none on the stalk which supports the solitary rosy pink flower, this orchid is readily distinguished from its closest relatives. It is often confused in conversation with *Cypripedium reginae* by many who also refer to the regal slipper as "pink."

Most cosmopolitan of all the orchids in its range, the Pink Lady's-Slipper stipulates only a high degree of acidity in its soil. It is found from the Arctic Circle to Tennessee in clay, sand, or even Jersey red soil; in swamps, on pine barrens, or on mountain tops. A pine grove or mixed evergreen forest is a favorite haunt, although beech and oak woods are by no means neglected.

In late May or early June these low growing orchids reach the peak of their short season of bloom. The two large, somewhat hairy, dark green leaves lie close to the ground and arise from a thick short underground stem sometimes mistaken for the rootstock. The three to five nerved leaves, sheathing the base of the flower stalk, are from six to eight inches long and two to three inches broad. They arch slightly so that their tips often touch the ground.

The flower, topping a slender stalk eight to twelve inches high, has a unique pouch. A long anterior fissure from the orifice to the base of the slipper divides the front of the rose-veined pink lip, but so ingeniously are the edges rolled inward that the length of the opening is not apparent to the casual observer, who assumes it to be but a fold in the moccasin.

This elongated opening makes access to nectar easy for visiting bees, but their attempts to retreat through the generous portal are met with the same rebuff they encounter in other species. In this Lady's-Slipper, as in the Larger Yellow, the surface of the stigma is beset with minute, rigid, sharp-pointed papillae, all directed forward, which are excellently adapted to brush off the pollen from the insect's head or back.

The pollen of the Pink Lady's-Slipper is more granular and less viscid than in other American species. The stigma is also viscid and slightly concave. Darwin could not find nectar within the lip. The tips of the hairs which line its inner surface, however, secrete minute drops of slightly sticky fluid. If sweet or nutritious, these would suffice to attract insects. It is certain that small bees frequently enter the pouch. The opening is too small for bumble bees, which find the sac an escape-proof trap.

The roughened pouch, often two inches in length, pendulous, and tipped slightly forward, is flanked on either side by a lance-shaped brownish petal slightly over an inch long. Two similarly colored sepals unite beneath the pouch, while a third forms the usual banner-like hood above it.

PINK LADY'S-SLIPPER
Cypripedium acaule

JOSHUA MAYER

PINK LADY'S-SLIPPER
Cypripedium acaule

SASATA

RAM'S HEAD LADY'S-SLIPPER
Cypripedium arietinum R. Br.

STATUS Michigan Special Concern

Although this orchid is known to be pollinated by bees, it reproduces mainly via offshoots from its rhizome, and plants may form small colonies.

The **Ram's Head** is the smallest of the lady's-slippers and is one of the rarest of our wild orchids. It has probably been seen growing in its illusive native haunts by fewer people than any other of our orchid species. Many botanists who have a specimen of the Ram's Head in their collections have never found it growing, although they may have searched diligently and hopefully for years. This is due in part to the fact that this is the most restricted in its distribution of any of the eastern North American lady's-slippers, and in part to the smallness and relative inconspicuousness of the flowers.

Ram's Head Lady's-Slipper grows from Maine to Minnesota, and into Saskatchewan. Fond of a cool climate and the shelter of evergreens, these small moccasin flowers bloom in May in cedar swamps, cold damp coniferous forests, on dry sandy hillsides beneath pines, and on evergreen heaths. Here, scattered around the base of juniper, balsam, or pine, and so completely concealed by overhanging branches that one must lift the lowest boughs to find them, the diminutive flowers grow singly at the summit of slender leafy stems from six to twelve inches tall.

The three sepals are greenish-bronze and, unlike those of all other lady's-slippers, are entirely separate from each other. This fact led some earlier botanists to place the Ram's Head in a group distinct from other *Cypripediums*. The upper sepal is elliptic, while the lateral pair and the petals are fringe-like and curled. The funnel-form pouch, which is only about an inch long and blunt at the apex, is whitish with deep magenta veins following the contour of the lip, whose rounded orifice is covered with hoary white pubescence.

The side view of this odd little pouch and the curving petals and sepals resemble very much the head and horns of a ram. The three or four smooth sessile clasping leaves are slightly bluish-green. They are bluntly elliptical, two to three inches long, and about an inch wide at their widest point. The rhizome or rootstock has a musky odor. The seed pod is an inflated brown capsule with prominent ridges.

RAM'S HEAD LADY'S-SLIPPER
Cypripedium arietinum
SUPERIOR NF

SMALL WHITE LADY'S-SLIPPER
Cypripedium candidum Muhl. ex Willd.

STATUS Michigan Threatened

The **Small White Lady's-Slipper** shares with the Ram's Head the distinction of being the smallest of the *Cypripediums*. The Small White slipper, however, is unique in its requirement of alkaline soil. Open marly bogs are its favorite haunts, where it grows equally as well in direct sunlight as in partial shade. It has the further distinction of being the only *Cypripedium* to grow on the open lands of the western prairie.

This early blooming, single flowered orchid is less than a foot high. The slender round stem is closely wrapped by the three or four crowded, erect, rather stiff, narrow leaves. Beset with minute bristles which are more abundant on the lower surface than on the upper, these prominently veined, lance-shaped leaves do not exceed five inches in length and are but an inch at their greatest width.

The waxy, dazzling white pouch, with its delicate violet or wine tinted veins within, is lined with long silky hairs. The pouch itself is less than an inch long and has a small horizontal orifice. The sepals are longer than the lip, and the two lower ones are completely united beneath it. The narrow, wavy, twisted petals are even longer than the sepals and are greenish-yellow spotted with purple. The rhizome, like the rest of the orchid, is small.

Where *Cypripedium candidum* and *C. parviflorum* are found growing closely together, hybrids are likely to occur; the hybrid between *C. candidum* and *C. parviflorum* var. *makasin* is known as *Cypripedium* ×*andrewsii*; the flower color may be white, creamy or yellow. This hybrid is uncommon in Michigan, being reported from several locations in the southwestern Lower Peninsula.

SMALL WHITE LADY'S-SLIPPER
Cypripedium candidum

MASON BROCK

SMALLER YELLOW LADY'S-SLIPPER
Cypripedium parviflorum var. *makasin* (Farw.) Sheviak

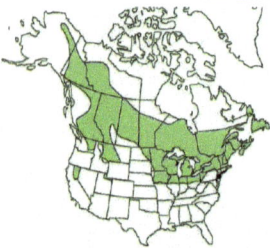

Cypripedium parviflorum (with its several varieties) is the most common wild orchid in North America, found across Canada and in nearly every U.S. state.

Many find it difficult to distinguish the **Smaller Yellow Lady's-Slipper** from the Larger. They may hybridize with each other to such an extent that one is frequently not certain his find is a pure strain of either. In determining either species one must inevitably refer to the other for comparison and for contrasting characters.

The Smaller Yellow Lady's-Slipper is at its best in swampy boggy situations, although it is also found on moist hilly woodlands. At the summit of a slender, minutely pubescent, leafy stem from one to two feet high, are the one or sometimes two smooth, glossy, golden yellow flowers.

Rarely more than an inch long, the tiny pouches hold a decided vanilla-like fragrance. On the back of the pouch Trelease described a variable number, one to four, of crescent-shaped or irregular translucent spots which readily attract the pollinating bee and lead it back under the stigma, from which point it can see the two small openings beneath the anthers through which to make its exit. In Trelease's experiments the small bees which he introduced into the lip usually went directly to the translucent spots, but finding no egress there went on to the regular exits.

Variously marked with streaks and flecks of rich wine-red within, the deep gold of the exterior finds a perfect foil in the claret of the long, narrow, twisted petals. The upper sepal, broadly ovate in form, makes a canopy over the lip, while the two lower ones are united beneath it. Only their ends are separate, giving the appearance of one sepal with a forked tip.

The leaves of the Smaller Yellow Lady's-Slipper are broader, somewhat longer, and have more pointed tips than those of the Ram's Head, but they are similarly ribbed with prominent veins, and clasp the stem in the same manner. They are from three to six inches long and about two inches

wide across the center. Their margins are smooth, but they are folded around the stem in such wise that their edges appear wavy or undulating. The ribbed brown capsule is 1½ to two inches long.

The brown rhizomes with tufts of fibrous rootlets break with a white fracture. Heavily scented like Valerian, these rootstocks yield the drug cypripedium, which was previously used as a nerve stimulant.

The Smaller Yellow Lady's-Slipper is a native of North America and has been cultivated since 1759. It is less plentiful than the Larger Yellow one, but it has a wider range and grows from Newfoundland to the Rockies. In Michigan it is most abundant in the Lower Peninsula. It blooms in May and early June.

SMALLER YELLOW LADY'S-SLIPPER
Cypripedium parviflorum var. *makasin*
DOUG MCGRADY

LARGER YELLOW LADY'S-SLIPPER
Cypripedium parviflorum **var.** *pubescens* (Willd.) Knight

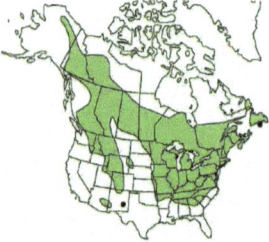

As its name implies, the **Larger Yellow Lady's-Slipper** exceeds in size the species just described. The entire plant is stouter, but is often very little taller. It ranges from 12 to 24 inches in height.

The clasping, broadly oval, pointed leaves are about half as wide as long, and usually attain a length of five or six inches. They have from seven to nine conspicuous parallel veins, and while entire the margins appear wavy. As in our other lady's-slippers, the glandular hairs of the plant's stems and leaves may cause skin poisoning if handled.

The rhizomes are rather stout and cylindrical and contain oil, resin, and tannin. They have a heavy odor and when manufactured into a drug the powder, infusion, or extract is bitter-sweet and pungent. Like many other Native American remedies, it is now little used.

The flowers are large, showy, and dull yellow, spotted and streaked within with madder-purple. They are somewhat coarse in texture. The lip is 1½ to 3 inches long, not glossy but slightly roughened on the surface, and very convex on top, where a tuft of white jointed hairs lies just within. The two linear, wavy, lateral petals are longer than the lip and are greenish-brown like the sepals. These, too, are long and lanceolate, and the two upper ones are partially united. The upper one bends slightly over the lip. The brown capsules contain minute elongated seeds with very thin coats.

Inside the lip, a triangular sterile stamen covering the stigma helps to assure fertilization, for by its position it effectively blocks the opening by which the bee enters and forces it to make its escape by one of the smaller exits at the rear, to reach which it must successively brush past the stigma and a pollen-bearing anther.

This orchid, like the Smaller Yellow Lady's-Slipper, is native to North America. It blooms in May and June in wetlands (especially those that are rich in calcium) and in all but the driest woodlands, and ranges across the continent.

LARGER YELLOW LADY'S-SLIPPER
Cypripedium parviflorum var. pubescens

ORCHI

SHOWY LADY'S-SLIPPER
Cypripedium reginae Walt.

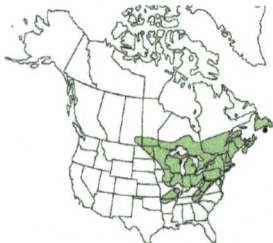

This superb orchid, known to many as the Queen Lady's-Slipper, is unquestionably the most regal of the *Cypripediums*. It inhabits places that are slow to warm up in spring, such as cool deep swamps of cedar and tamarack; spruce and sphagnum bogs with open, dark, evergreen-fringed pools; but it also grows in great luxuriance in marshy open glades bordering mucky tamarack-edged lakes in company with gray dogwood, poison sumac, and shrubby cinquefoil.

Even when growing in lush profusion with acid loving cranberries, labrador tea, and laurel in the heart of a sphagnum bog, close examination will reveal that the Showy slippers but tolerate the acidity and that their fleshy scarred rhizomes and fibrous roots grow in a neutral or slightly alkaline plane.

This Lady's-Slipper blossoms later than the others, opening its flowers when the Yellow and Pink species are nearly at the end of their period of bloom, and long after the Ram's Head and Small White have completely faded. One may look for it during June and July in Michigan. The farther north it grows the later it comes into bloom.

Showy Lady's-Slipper grows slowly, taking as much as 15 years to produce its first bloom; plants can live over fifty years. Under favorable conditions, a single plant can produce over 200 flowering stems.

Outstanding in every respect, the Showy lady's-slippers are the tallest of their kin. The stout almost bristly stem, which bears five to seven large, closely enfolding hairy leaves, is from two to three feet high. These five to seven leaves are broadly ovate and appear crowded on the stem to a point within a few inches below its summit, where, more clearly visible, the stem becomes more slender, has softer hairs, and at length reaches the culmination of its purpose in bearing from one to four flowers of unparalleled beauty.

Fully distended yet fragile and shell-like, the large milk-white pouches are an inch deep, an inch across, and more than an inch long. Purple markings within show through its transparent walls, causing one writer to describe it as a "large white sac, with splashes of purplish-pink looking as though wine had overflowed its cup and trickled down the sides." On each side is a narrow white petal, while beneath it two greenish-white sepals unite. Above the pouch and slightly inclined over it, yet broadly oval and banner-like, is the third sepal, which is also greenish-white. The margin of the pouch is slightly drooping, conspicuously flattened, and its edges deeply inflected. A heavy hairy ridge within, extending from the column to the orifice, serves to prevent the escape of insects through the large opening and forces them to emerge as in other lady's-slippers through the smaller tradesmen's exits at the rear.

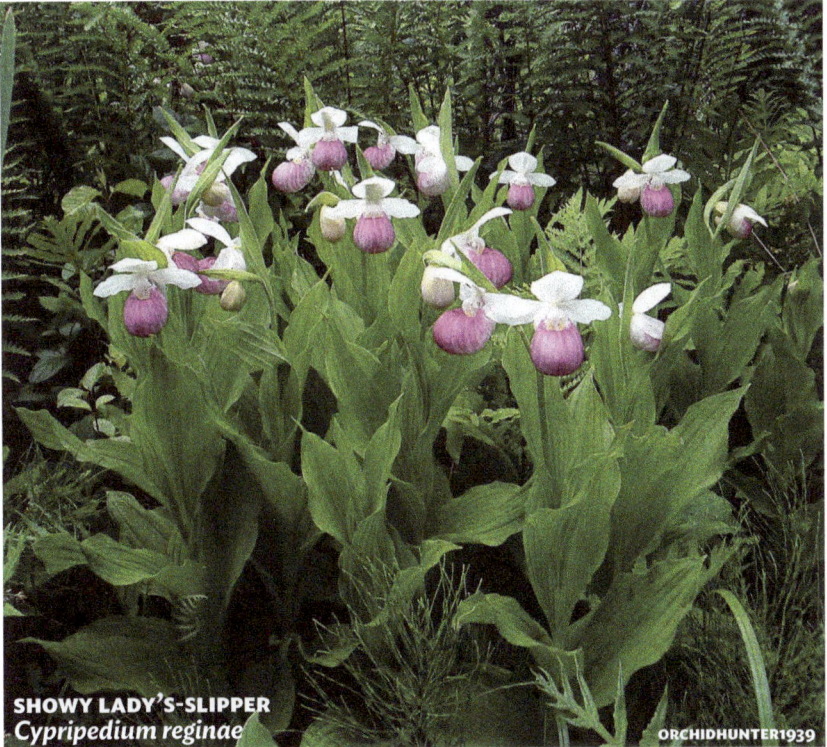

SHOWY LADY'S-SLIPPER
Cypripedium reginae

ORCHIDHUNTER1939

SHOWY LADY'S-SLIPPER
Cypripedium reginae

ORCHI

LONG-BRACTED ORCHID

Dactylorhiza viridis (L.) R.M. Bateman, Pridgeon & M.W. Chase

SYNONYMS *Habenaria viridis* (L.) R. Br., *Coeloglossum viride* (L.) Hartman

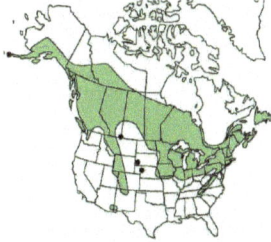

The **Long-bracted Orchid**, traditionally grouped with *Habenaria* and later as a *Coeloglossum*, deserves more attention than it ordinarily receives. Summarily dismissed because of its inconspicuous green flowers, it is one of the most graceful of all wild orchids. Fond of moist woods and meadows, it blooms from May to August in sub-acid soil. While it may, as Morris and Eames believe, prefer hardwoods and particularly upland woods of beech and maple, it also thrives in beech-evergreen forests in the Great Lakes region, and is associated with the Large Round-leaved Orchid and the rattlesnake plantains.

A tall plant from six to twenty-four inches in height, this orchid, like the lady's-slippers, has a leafy stem. Its smooth lance-shaped leaves, however, have the distinguishing character of being largest at the base of the stem and diminishing in size as they go upward. The lowest ones are five to six inches long and are broadly ovate; but the higher up they grow, the shorter and more slender they become, until those nearest to the spike of flowers are reduced to mere bracts. From these long floral bracts and greenish flowers this orchid receives its name (another oft-used name is Frog Orchid).

Long-Bracted Orchid has one of the widest global distributions of any orchid: it is found across Eurasia, Canada, and the United States, from Alaska to North Carolina.

The individual flowers are small, scarcely a third of an inch long; but arranged in a raceme from two to six inches long, each flower subtended by a slim bract two or three times as long as itself, the cluster of blossoms is arresting. The sepals and petals are separate; the sepals are green and the petals greenish-white. The greenish-white, two-lobed lip, with a small tooth between the lobes is a determining character. Although the lip itself is scarcely a third of an inch long, the white sac-like spur is but half that length.

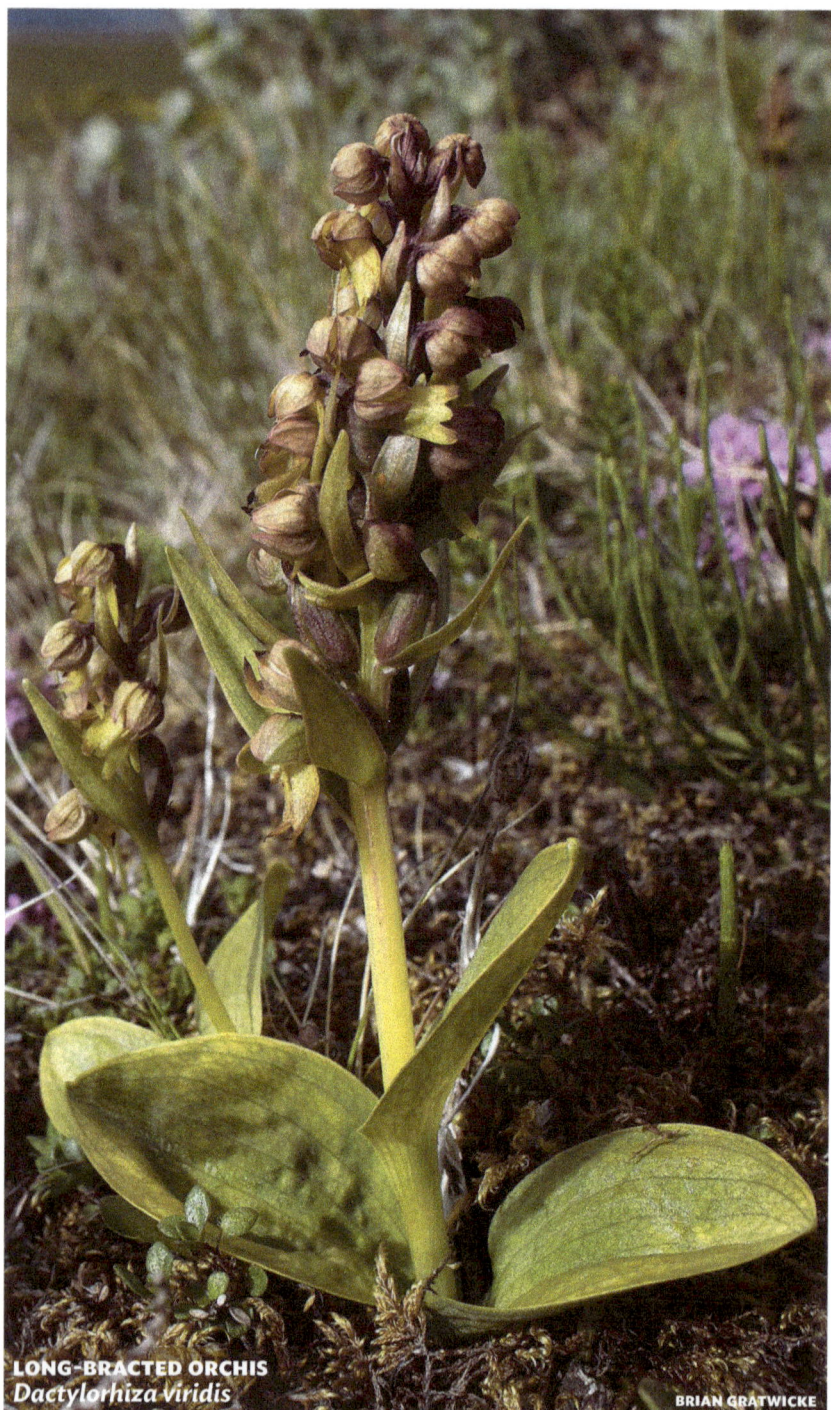

LONG-BRACTED ORCHIS
Dactylorhiza viridis

BRIAN GRATWICKE

HELLEBORINE
Epipactis

EPIPACTIS IS A GENUS OF ABOUT 75 SPECIES found in North and Central America, Eurasia, and North Africa. There are two species in North America, one native to the western states (**Giant Helleborine**, *Epipactis gigantea*), and *Epipactis helleborine*, introduced from Europe, and first reported from Syracuse, New York in 1879, where it had been planted by European immigrants, presumably for medicinal purposes. Today, Helleborine is widely naturalized in the northeast states, the Great Lakes region, and to a lesser extent, along the west coast. It has continued to spread to the point where it is sometimes called the "weed orchid."

HELLEBORINE
Epipactis helleborine (L.) Crantz

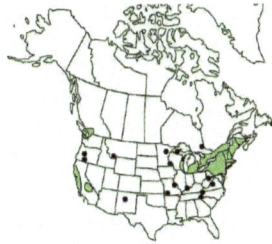

First recorded in Michigan in 1919 in Berrien County, today **Helleborine** is the state's most common orchid (although rarely forming large colonies), and continues to spread. Its favored habitats are in the shade of deciduous forests and mixed hardwood-conifer forests, but it is also at home in gardens and yards.

Flowering in late summer and early fall, plants of Helleborine may grow two to three feet tall, with three to ten leaves on its finely hairy stems. Numerous small flowers, 15 to 30 (up to 50) in number, grow from a loosely one-sided raceme. The flowers have greenish purple petals and sepals, and a lip divided into two segments by a central constriction; the innermost part of the lip, closest to the column, is curved into a bowl-like shape, is purple or brown, and often glossy on the inner side; the outer side is pink, green or white. The roots are fibrous.

HELLEBORINE
Epipactis helleborine

CBAILE19

SHOWY ORCHID
Galearis

ALTHOUGH THERE ARE ABOUT EIGHTY SPECIES of *Galearis* in the north temperate zone, only two are found in North America; both grow in Michigan. Our two species are easily recognized by their low racemes of spurred pink or rosy purple and white flowers rising from one or two large leaves. Their roots are fleshy tubercles.

SMALL ROUND-LEAVED ORCHID
Galearis rotundifolia (Banks ex Pursh) R.M. Bateman

SYNONYMS *Amerorchis rotundifolia* (Banks ex Pursh) Hultén, *Orchis rotundifolia* Banks

The single, roundish basal leaf serves well to distinguish this orchid from its more widely known sister, *Galearis spectabilis*. Lover of cool northern cedar and tamarack swamps (and where often underlain by calcium-rich marl), the round-leaved orchis favors very moist, shaded, mossy pockets of sphagnum and other wetland mosses. From early June until early July those who search diligently may find this rare orchis in the vicinity of the Straits of Mackinac and in the Upper Peninsula.

A slender leafless stalk, six to ten inches high, bears a loose raceme of a few small delicately-tinted flowers. The three pale pinkish-mauve sepals are slightly longer, wider, and deeper in tint than the petals. The white lip is conspicuously spotted with purple and cleft into three lobes, the middle one of which is forked. The spur is short and slender.

The solitary dull green leaf, but three inches long and half as wide, has earned for this orchid the another common name of One-leaf Orchid.

SMALL ROUND-LEAVED ORCHIS
Galearis rotundifolia

NC ORCHID

SHOWY ORCHID
Galearis spectabilis (L.) Raf.

SYNONYM *Orchis spectabilis* L.
STATUS Michigan Threatened

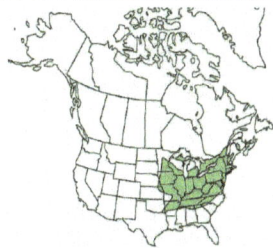

The genus name, *Galearis,* is derived from the Latin word, *galea,* meaning helmet. It refers to the two pink to purple upper petals which form a hood over the flower.

The **Showy Orchid**, in contrast to the round-leaved species, grows in the deciduous forests of Michigan's Lower Peninsula. Rich, moist (but well-drained) woodlands are its home, where moderate shade permits the growth of its frequent companions—hepaticas, trilliums, and dog's-tooth violets. It also does well in somewhat disturbed places as along trails or in the flood zone of small streams.

The Showy Orchid is one of the earliest flowering members of its entire family, for it appears soon after bloodroots in early May and continues in bloom for about a month. This well-known orchid, although scarcely a foot high, is conspicuous with its three to eight large mauve and white flowers. They are fully an inch long and grow in a loose spike, each of them accompanied by a long green bract. The petals and sepals completely fused together form a forward arching hood which bends over the column.

The pure white lip, three-quarters of an inch in length, is tongue-shaped and deflexed. It extends backward into a long spur containing nectar to attract pollinating bumble-bees. The mouth of the spur, just wide enough to admit the bee's head, is a special adaptation to secure cross fertilization. As the head of the visitor goes into the nectary, it brushes past the pollen masses, ruptures the thin membrane which covers them, and immediately exposes two sticky discs which at once become affixed to the head of the bee. The orchid's pollen masses are, of course, attached to the discs. When the insect enters another orchis, these discs and pollen masses are in the exactly correct position to become attached to that flower's sticky stigma.

Although the blossoms of Showy Orchid are delicately beautiful, the entire plant is somewhat fleshy in appearance. At the base of the stout, four or five angled stem are two large, flat, smooth, light-green, shining leaves three to six inches long. They are oblong in shape, tapering to a point at the tip and gradually narrowing to a groove at their basal end so that they quite envelop the stem. The roots are oblong and fibrous.

SHOWY ORCHIS
Galearis spectabilis

ANTEPENULTIMATE

RATTLESNAKE PLANTAIN
Goodyera

FOUR OF THE NEARLY 100 SPECIES of rattlesnake plantains in the world may be found in North America, and all four are found in Michigan. While occasionally in the more southern hardwood forest regions, they distinctly prefer the cool shelter of northern pines, hemlocks and cedars.

Their rosettes of velvety gray-green leaves with white markings set them apart. These markings, which resemble those on a snake, are responsible for the name, but the Indians also once thought these plants to have curative powers for the bite of a rattlesnake. Because the leaf rosettes are evergreen and so ornamental, and because they will thrive in small quarters under glass, various species of rattlesnake plantain are often used in terraria.

The numerous small, white, round-bodied, spurless flowers are borne in spikes. The base of the lip is curled into a sac with its edges turned outward. Although not infrequent in the regions where they occur, they are not free-blooming, and one may find but 20 plants in a 100 bearing flowers. Regardless of their prodigality in producing seed, rattlesnake plantains assure their continuity as their thick, fleshy fibrous roots creep underground and send up new plants at intervals. This habit of growth accounts for the fact that one generally finds these orchids in colonies.

Petals and upper sepal unite to form a hood while the lateral sepals are wide-flung and the lip is rounded into a sac. The flowers, supported by inconspicuous bracts, grow in a compact spike.

Like other members of its family, rattlesnake plantains have specific means for securing cross pollination. The minutely-toothed sticky stigma is overshadowed by the column, which projects up and over it and holds the anther on a little extended arm. Behind the arm is a small cup-shaped cavity holding two pollen masses which are united at their base to a common sticky gland. When bumblebees seek nectar within the sac-like lip they, in thrusting their tongues through the narrow opening, must pass the tip of the anther disc. As the insect withdraws, the membrane covering the pollen masses is ruptured and the sticky disc with the pollen masses adheres to the insect's tongue. When it enters a new flower, the bee leaves its burden on the receptive under side of the stigma.

The scientific name honors English botanist John Goodyer (1592-1664).

GREEN-LEAF RATTLESNAKE PLANTAIN
Goodyera oblongifolia Raf.

SYNONYM *Goodyera decipiens* (Hook.) F.T. Hubb.

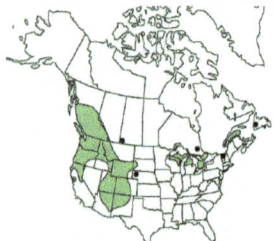

Our only **Goodyera** lacking the conspicuous net-like venation on its leaves; typically, only the midvein of the leaf's upper surface is outlined in white or pale green.

Largest of all the rattlesnake plantains and mostly confined to dry evergreen forests (rarely under deciduous trees), the leaves of this species at once set it apart from the other three members of its genus: they are but rarely mottled, often entirely devoid of white markings, but generally have a broad stripe down the mid-rib. Occasionally the median white stripe branches, feather-fashion, the lines becoming fainter the farther they are from the mid-rib.

Though velvety and dark green, the leaves have less of the blue-green tone than those of the other rattlesnake plantains. They are rather stiff, lance-shaped, from two to four inches long, taper at both ends, and have wavy or fluted margins.

The tall, stiff, fourteen to eighteen-inch stem is somewhat hairy and terminates in a four-inch long, one-sided tapering spike of greenish-white flowers. The individual blooms are larger than those of the other species and average from ¼ to ⅜ inch in length, nearly half of which is occupied by the elongated lip that, in this rattlesnake plantain, is large at the base rather than sac-like. It is centrally grooved and tapers to a blunt tip. The edges of the lip are curved inward providing a well defined channel to the nectary.

Green-Leaf Rattlesnake Plantain blooms in August, and although transcontinental in its range, it is more abundant from Lake Huron westward than in the eastern United States, where it is known only from Maine. In Michigan, plants occur from the upper limits of Saginaw Bay northward.

MENZIES' RATTLESNAKE PLANTAIN
Goodyera oblongifolia

DOWNY RATTLESNAKE PLANTAIN
Goodyera pubescens (Willd.) R. Br.

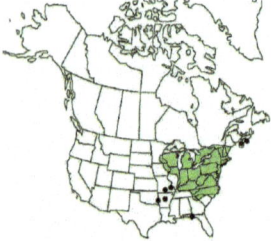

Best known of the rattlesnake plantains, the Downy blooms in August in dry upland forests of conifers or mixed conifer-deciduous woods. Soils are often sandy and somewhat acidic.

Identified by its rosette of bluish-green leaves with their network of intricate and fine white markings, this species has flowers which differ from the others of its group. Although growing in the usual spike, the individual white blooms are distinctly different in having a rounded contour. Petals and sepals contribute to the spherical outline by being unmistakably globular though their outer edges spread slightly. The lip is fully inflated.

The general impression of roundness is furthered by the stout, cylindrical, blunt-topped spike in which the flowers are arranged in a true but densely compacted spiral.

The downy stem rises from oval-leaved evergreen rosettes and varies from six to 16 inches in height.

The thick, perennial underground rootstock assures perpetuation of this hardy species which, like our other *Goodyera,* fails to provide all of its plants with flowers each season.

Downy rattlesnake-plantain is the commonest *Goodyera* in the southern half of the Lower Peninsula, and the only one in Michigan's southernmost counties; in the Upper Peninsula, it is more local.

DOUG MCGRADY

JOSHUA MAYER

DOWNY RATTLESNAKE PLANTAIN
Goodyera pubescens

JOSHUA MAYER

LESSER RATTLESNAKE PLANTAIN
Goodyera repens (L.) R. Br.

A woodland species, the **Lesser Rattlesnake Plantain** prefers the cool shade of evergreens. It is the smallest of the rattlesnake plantains and the most widely distributed, completely circling the globe in the northern hemisphere.

Recognized on sight chiefly by the conspicuously white bordered veins of its rosette of small blue-green leaves, this species is unmistakably identified upon closer observation by its low, one-sided, slightly woolly spike of small greenish-white flowers whose lips are sac-shaped, taper to a sharp point, and abruptly bend downward toward the apex. Fuller describes the characteristic veining of the leaf which distinguishes it from other species: "All save the midrib are conspicuously bordered with white. The comparatively few lateral veins join the longitudinal ones at about right angles and are unbranched."

Typically in moderately acid soil of drier evergreen or mixed woods (or sometimes in hummock in cedar swamps), this orchid blooms in July and August, sending up its floral stalks to a height of six to nine inches. The individual flowers, ⅛ inch long, are each subtended by a bract. From ten to twenty flowers comprise the spike.

UMBERTO FERRANDO

LESSER RATTLESNAKE PLANTAIN
Goodyera repens

JOSHUA MAYER

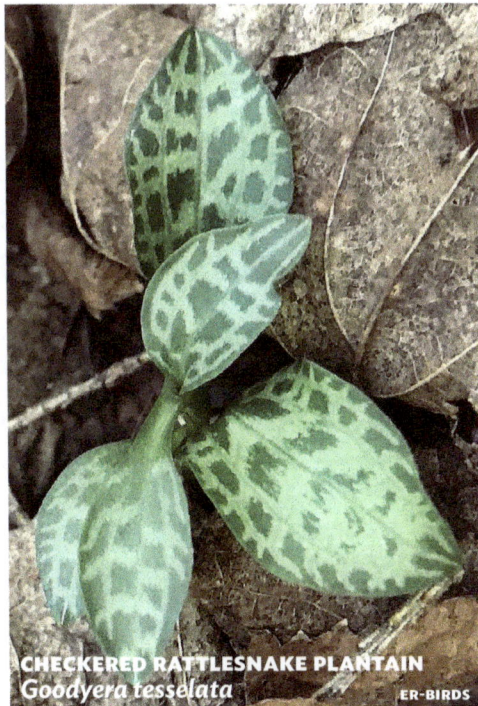

CHECKERED RATTLESNAKE PLANTAIN
Goodyera tesselata Lodd.

Taller, stouter, and with flowers nearly twice the size of those of the Lesser Rattlesnake Plantain, this species, though less widely distributed, is often found growing and blooming simultaneously in the same evergreen forests and swamps as its smaller relative; also known from hollows in sand dune areas.

Unlike the Lesser Rattlesnake Plantain, the flowers of Checkered are not arranged in a one-sided spike but in a cylindrical or spiralled raceme. The lip has a less pronounced sac with a blunt rather than a sharply pointed tip. The five- to nine-veined, oval to lance-shaped leaves are pencilled with pale green or white, and often grayish-green, in contrast to the darker blue-green leaves of Goodyera repens.

Checkered rattlesnake plantain often grows together with *Goodyera oblongifolia* or *G. repens,* and apparently occasionally hybridizes with them.

CHECKERED RATTLESNAKE PLANTAIN
Goodyera tesselata

DOUG MCGRADY

ER-BIRDS

WHORLED POGONIA
Isotria

Closely related to **Nodding Pogonia** (*Triphora trianthophoros*), the two known species of this group have distinctions sufficiently well-marked to place them in a separate genus. They are low plants which rise from a rootstock and bear one or two flowers at the apex of their stems. The lip of the flower is crested but has no spur. The leaves are whorled.

Two species are reported for Michigan, but the **Smaller Whorled Pogonia**, *Isotria medeoloides* (Pursh) Raf., first discovered in 1968 in Berrien County in the extreme southwest Lower Peninsula, is likely no longer present in the state, and is not described here (federally listed as Threatened).

WHORLED POGONIA
Isotria verticillata (Muhl. ex Willd.) Raf.

STATUS Michigan Threatened

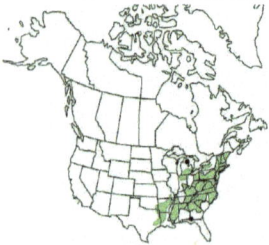

The **Whorled Pogonia** may be easily distinguished from Nodding Pogonia by the arrangement of its leaves, which are whorled. The circle of five or six blue-green stemless leaves, each two to three inches long, is perhaps six inches in diameter when fully mature. The leaves are about half grown when the orchid is in bloom, and although they are broadest near the outer end, they have abrupt tips.

Because of their arrangement and the situations and companions they choose, Whorled Pogonia is sometimes mistaken for the closely resembling Indian cucumber-root (*Medeola virginiana*). From this associated plant it can be distinguished by the absence of a colored spot which centers the leaves of Indian cucumber-root. This latter plant sometimes bears two whorls of leaves, the Whorled Pogonia but one; the cucumber root's stems are wiry, those of pogonia pale purple, plump, translucent, hollow, and almost succulent.

From the axil of the leaf-whorl rises a solitary flower or, infrequently, two flowers, each borne at the summit of a slender inch-long stalk which, while it supports the flower, bends forward. When the capsule is ripe, however, it becomes erect and a full inch and three-quarters long.

The flower is unique. Its two-inch long, narrow, strap-like, spreading sepals of dusky purple are twice as long as the greenish-yellow petals which, lance-shaped, arch together over the lip. The wavy-margined lip, in turn, is shorter than the petals; it is three-lobed. Both lateral lobes are streaked with purple and curve upward; the middle one is prolonged into a wide rounded platform, down the center of which, as described by Morris and Eames, is a "flat-topped greenish-yellow ridge of waxy-looking material; at its base lie a pair of conspicuous orange spots and flanking it on each side a series of dull purple streaks."

This extraordinary orchid averages but ten inches in height, rising from a fleshy rootstock with several dark and fuzzy underground runners which have pale soft tips that spread with ease through the moss or soft humus in which the plant grows. Preferring very acid and moist soil, Whorled Pogonia is, notwithstanding, also sometimes found on dry hillsides, in pine-barrens, and in open dry woods, especially northward in Michigan. Moist woods and sphagnous swamps are its more typical habitat. Shading of the roots, and sunshine for the leaves and floral parts are preferred.

Through May and June Whorled Pogonia blooms east of the Mississippi River and south of the Great Lakes. It is most plentiful in the more southern portions of its range. Although well distributed, it is unfamiliar to many because of the inconspicuous color of its flowers and the fact that it is frequently overshadowed by the laurels and azaleas among which it grows. It is probably most often associated with aromatic wintergreen (*Gaultheria procumbens*).

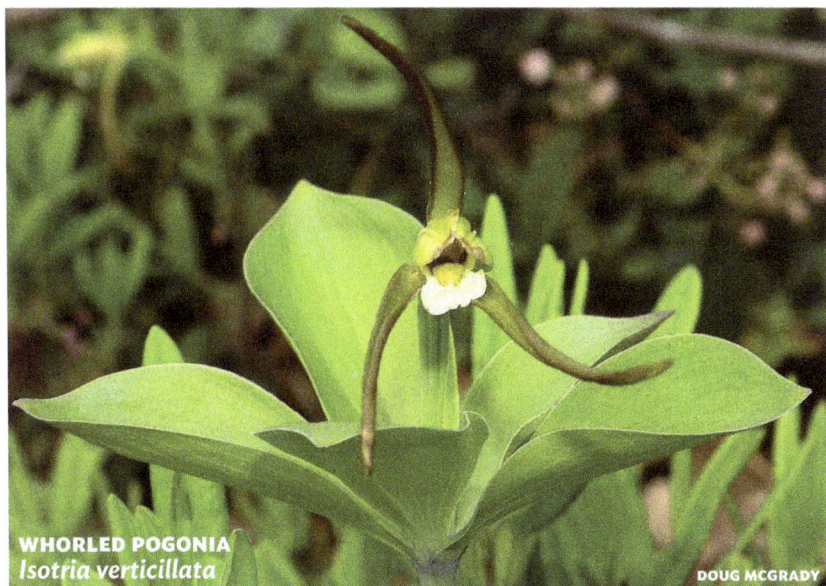

WHORLED POGONIA
Isotria verticillata

DOUG MCGRADY

WHORLED POGONIA
Isotria verticillata

JASON HOLLINGER

TWAYBLADE
Liparis

THE COMMON NAME OF THIS GROUP of orchids is somewhat misleading since it corresponds with that of another previously described (*Neottia*). The scientific names, of course, differentiate them, as do the characteristics of the plants themselves.

Liparis, a Greek derivative, means "shining" and refers to the appearance of the two roundish or lance-shaped basal leaves. The plant grows from a tuber and bears its slender petaled flowers in a spreading raceme.

Two large, waxy pollen masses are produced in each anther cell. The expanded lip has a nectar-secreting groove down its median line which with its incurved edges provides a channel for the visiting insect, leading toward the stigma and anther cells.

The two twayblades found in the northeastern states are present in Michigan.

LARGE TWAYBLADE
Liparis liliifolia (L.) Rich. ex Lindl.

This woodland orchid is perhaps better known by its other common name, **Lily-leaved Twayblade**, which is a literal translation of its specific name from the Latin.

In open upland woods, growing in light sandy moderately acid soil, the Large Twayblade blooms in June and July. It also does well in areas once cultivated or disturbed such as second-growth woods and pine plantations.

Plants are rarely more than seven inches tall and so inconspicuously greenish watery-mauve that one can be fairly upon it before becoming aware of its presence— a foot or two away, it blends so perfectly into its surroundings that it cannot be seen.

The two glossy, light green, oval leaves clasp and sheath the stout angled stem at its base. The mauve-green flowers of the raceme each have a stalk so that they are individually apparent. This somewhat divergent habit of the blooms gives the flowering spike a diameter of fully an inch.

Each flower is about a ½ inch long. The greenish-white sepals are lance-shaped, but their edges are rolled inward so that they appear thread-like. The

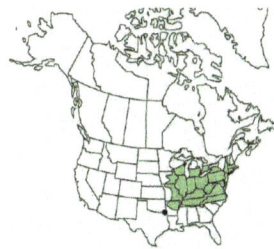

upper sepal curves backward, the lateral ones downward. The greenish-mauve petals are also thread-like, for they are twirled into long hollow tubes and bend down but have their tips pointing forward. The vari-colored lip, mauve, purple, and green, is wedge-shaped with clasping lobes at its base. Its expanded portion is widely rounded and oval, broadest at its outer edge. Through the median line is a glassy ridge which projects over the outer edge of the lip as a small tip. This central ridge is bordered on either side by a crimson finely branched vein.

The seed pod, a ribbed capsule, continues to be erect on the stalk as it ripens.

ALLEFANT

JASON HOLLISTER

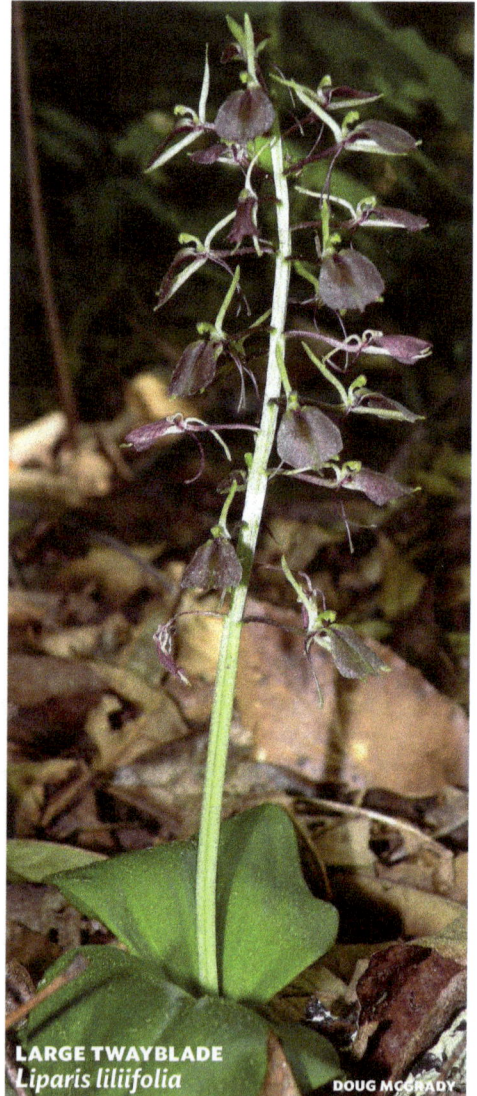

LARGE TWAYBLADE
Liparis liliifolia

DOUG MCGRADY

LOESEL'S TWAYBLADE
Liparis loeselii (L.) Rich.

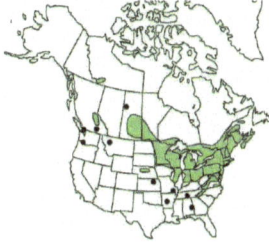

More northern in its range than *L. liliifolia,* **Loesel's Twayblade** blooms in May and June in bogs, wet woodlands, moist sandy meadows, and on wet forested banks and cliffs. Its yellowish-green floral organs, with an entire absence of mauve, are distinctly different from those of the Large Twayblade. The stem, however, is angled like that of its closest relative and rises similarly between two oval basal leaves.

A low-growing orchid, Loesel's Twayblade averages only eight inches in height. The upper third of the stem is a loose, few-flowered raceme of pale yellowish blossoms. Both petals and sepals are narrow and spreading. The lip is narrowly wedge-shaped and concave, turning upward at the inner end but downward at the outer. Its median ridge extends like that of the Large Twayblade into a small projecting tip.

In many respects these two so closely related species have entirely opposite characteristics. Loesel's Twayblade prefers a more northern habitat and decidedly moist situations rather than those that are better drained and more or less dry. This species is quite distinct in its coloring, and while many structural points are technically identical, the general contour of the

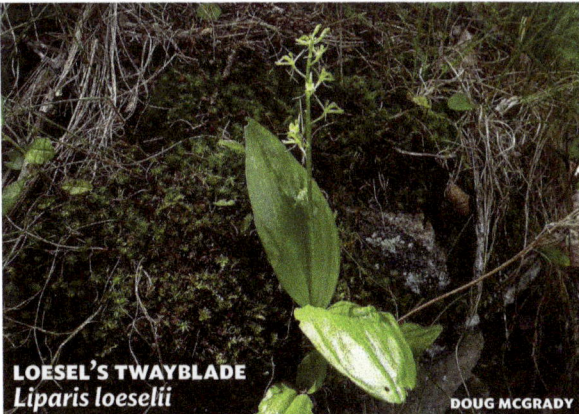

LOESEL'S TWAYBLADE
Liparis loeselii
DOUG MCGRADY

plants is dissimilar. The arrangement of the floral spikes contributes to this impression, for while both bear racemed narrow-petaled flowers, Loesel's Twayblade has fewer of them and the individual flowers are but half as long.

LOESEL'S TWAYBLADE
Liparis loeselii

ORCHI

ADDER'S-MOUTH
Malaxis

ABOUT 180 SPECIES OF ADDER'S-MOUTH grow throughout the world, mostly in the tropics; nine of them in North America, but only two are known in Michigan. The generic name signifies "smooth" and alludes to the tender texture of the leaves.

They are low-growing insignificant plants, rising but a few inches from a solid bulb. The slender stems are clasped a little above the ground by a single oval sheathing leaf and terminate in racemes of tiny white or greenish flowers. The efficiency of their devices to assure cross pollination by insects, in spite of their lack of odor and attractive color, is evident by the immediate appearance of rounded ripening seed capsules after the flowers have faded. The pollen is in four waxy masses, two in each anther cell. The sepals are strap shaped while the petals are so narrow that they are mere threads. The lip is slightly expanded.

Adder's-Mouth is not plentiful but it may be found in moist northern forests and in cool shaded bogs throughout much of Michigan. It quickly disappears with the clearing of the forest but may be discovered on old logging trails where new growth is replacing the cut timber.

WHITE ADDER'S-MOUTH
Malaxis monophyllos (L.) Sw.

Requiring situations that remain moist throughout the growing season, **White Adder's-Mouth** is found on low, damp, mossy floors of mixed forests and swamps; in the north it can also be seen on sandy dunes near Lake Superior. Soils are neutral, alkaline or only slightly acid.

No more than six inches high, this tiny orchid bears at the top of its slender stem a very compact spike about 2½ inches long containing from twenty to forty short-stemmed minute flowers, so small that the entire spike is but a quarter of an inch in diameter. The thread-like petals and sepals are greenish or watery-white and scarcely noticed by the unaided eye. The distinguishing lip, itself a mere 1/12 of an inch long, is widely rounded at its base, abruptly constricted in the middle and tapers on its outer end to a sharp point. The single oval leaf clasps the stem near the root.

White Adder's-Mouth blooms in June and July. It seems to prefer limestone regions and northern latitudes, but it has been found in such apparently contradictory areas as Kalamazoo and Washtenaw Counties in Michigan.

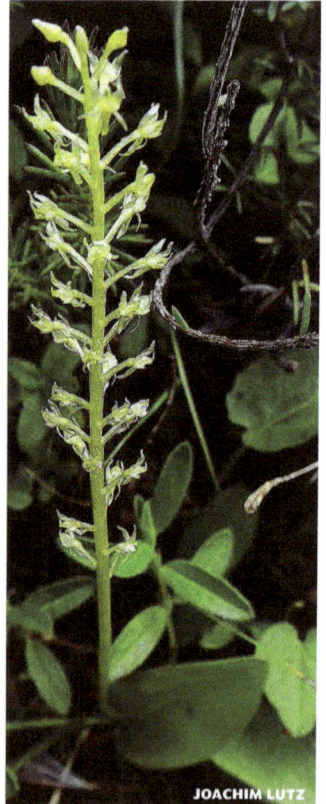

BERND HAYNOLD

WHITE ADDER'S-MOUTH
Malaxis monophyllos

BARTOSZ CUBER

JOACHIM LUTZ

GREEN ADDER'S-MOUTH
Malaxis unifolia Michx.

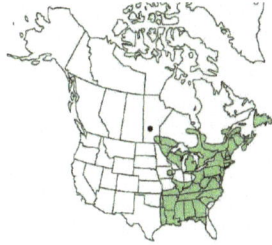

The **Green Adder's-Mouth** is slightly taller than the White and bears its single oval bright green leaf about midway up its smooth slender stem. The floral raceme, one to three inches in length, is fully an inch in diameter and consists of from twenty to sixty flowers, each with a slim spreading stalk one third of an inch long. It is these flower stalks which give breadth to the cluster and easily distinguish this species from the smaller species of Adder's-Mouth.

The green petals and sepals are narrow; the lip has a pair of backward pointing lobes at its base and at its outer end two large lobes with a small one between them. Fuller describes the peculiar change in position of the lip that occurs as the flowers mature. When the buds first open, the lip is below the petals and sepals, but as the flower becomes older it is entirely inverted and the lip, instead of the petals and sepals, is uppermost. Fuller also observes that few flowers are fertilized and that the unfertilized ones remain attached to the raceme for a long time.

Green Adder's-Mouth prefers very acid, typically sandy or rocky soil (often under bracken fern), and is found on dry wooded hillsides as well as in conifer swamps and bogs. It blooms during July and August.

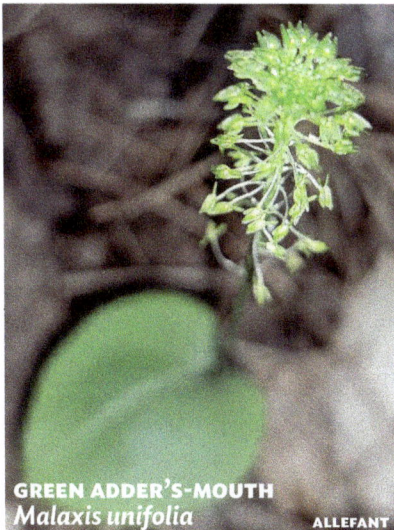

GREEN ADDER'S-MOUTH
Malaxis unifolia   ALLEFANT

TWAYBLADE
Neottia

THE TWAYBLADES ARE SMALL INCONSPICUOUS ORCHIDS which prefer a cool climate. Previously, our plants were known as *Listera* in honor of an English naturalist of the seventeenth century, Martin Lister.

Twayblades have fleshy fibrous roots, and, as the common name implies, a pair of small oval leaves midway up their short stems which terminate in a loose raceme of small greenish-purple flowers. The petals and sepals are much alike and the flowers have no spurs. The species are distinguished by the shape of the lip. Seeds are developed in a small ovoid capsule.

Darwin was intensely interested in the manner in which these orchids are fertilized and sent his son forth to watch long hours in an effort to discover what insects transferred the pollen and how they did it.

In these minute flowers the anther, with its pollen masses, is separated from the stigma by the thin leaf-like tip of the column, which is extremely sensitive and secretes a drop of sticky fluid the instant it is touched. At the same moment, the pollen masses are ejected so that the visiting bees and flies receive, simultaneously, adhesive fluid and pollen when they inevitably touch the column in their quest of the nectar so copiously provided along the ridge of the lip above its split end.

Three species of twayblades (and one hybrid) are found in Michigan; worldwide, nearly 70 species are described.

AURICLED TWAYBLADE
Neottia auriculata (Wiegand) Szlach.

SYNONYM *Listera auriculata* Wiegand
STATUS Michigan Special Concern

In Michigan, **Auricled Twayblade** is found on Isle Royale, the Keweenaw Peninsula, and in the eastern Upper Peninsula. It is most often seen in the sandy soil borders of swamps and in alder thickets; it is less often found in mixed woods or the spruce-fir forests of Isle Royale. It is a northern species becoming more common in eastern Canada.

No taller than the **Heart-leaved Twayblade**, this species has whitish-green flowers about twice as large, which are characterized chiefly by a pair of rounded ears at the base of their lips. The pale green stem is minutely downy above the two opposite, thin, smooth, oval leaves which, like the flowers, are

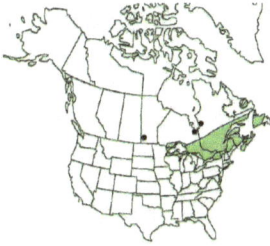

nearly double the size of those of its heart-leaved relative. Petals and sepals are all spreading, and while the split round-lobed lip is oblong and about equally as wide at one end as at the other, it is slightly pinched or constricted in the middle.

The Auricled Twayblade blooms in June and July and prefers at least moderately acid soil.

Hybrids with *Neottia convallarioides* are reported from Alger and Luce counties, and have been named *Neottia* ×*veltmanii* (Case) Baumbach.

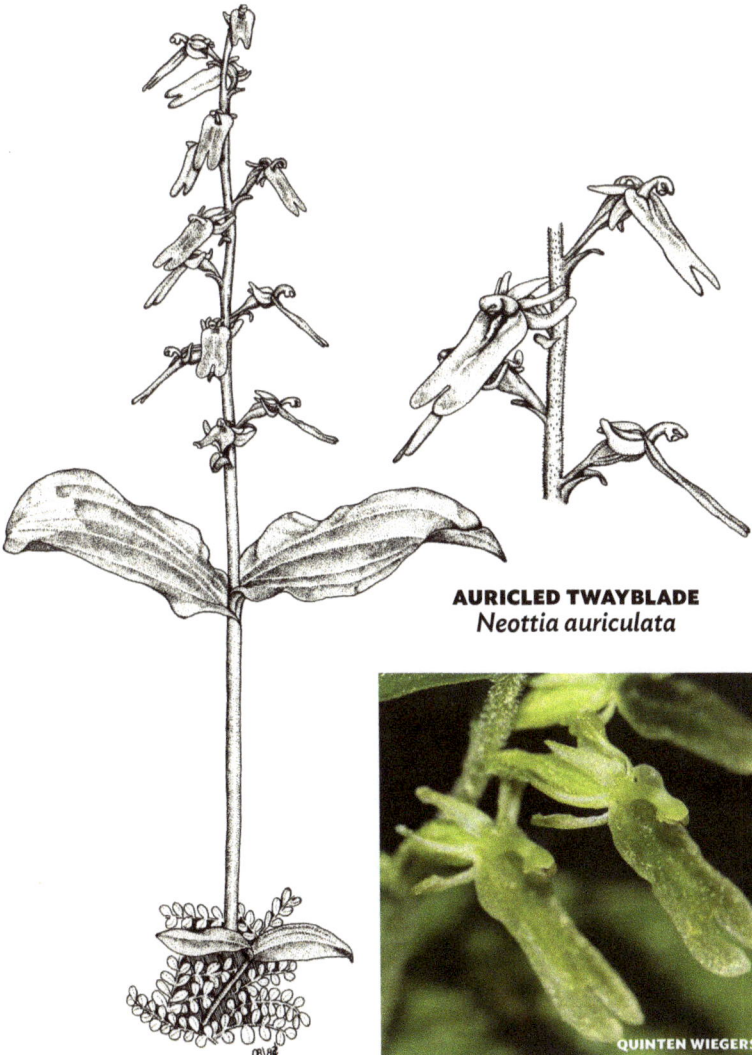

AURICLED TWAYBLADE
Neottia auriculata

QUINTEN WIEGERSMA

BROAD-LEAVED TWAYBLADE
Neottia convallarioides (Sw.) Rich.

SYNONYM *Listera convallarioides* (Swartz) Nutt.

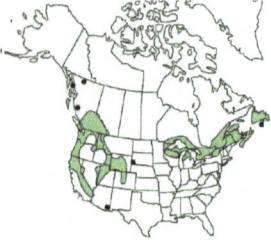

Named from a fancied rather than a real resemblance of its leaves to those of the Lily-of-the-Valley, this pale watery-green flowered twayblade blooms in June and July in rich, moist, leaf mold. It chooses wet situations, and while perhaps occurring more often in the shelter of deciduous trees, it also grows under evergreens, often in the wet sandy soil along a stream.

The stout stem is distinctly hairy above the two smooth roundish-oval leaves (which may not be exactly opposite each other). The upper third of the six to ten inch stem is occupied by a loose raceme of short-stalked, inconspicuously colored flowers. Their watery-green color is sometimes marked with purple. The wedge-shaped yellow-veined lip is widest at its outer edge, where it is shallowly cleft by a wide notch into two rounded lobes. A median glassy ridge projects just enough to form a minute third lobe in the apex of the notch. At its inner edge the lip is clawed.

Because the Broad-leaved Twayblade grows in colonies, it is more easily discovered than our other two species.

BROAD-LEAVED TWAYBLADE
Neottia convallarioides

RYLAN SPRAGUE

HEART-LEAVED TWAYBLADE
Neottia cordata (L.) Rich.

SYNONYM *Listera cordata* (L.) R. Br.

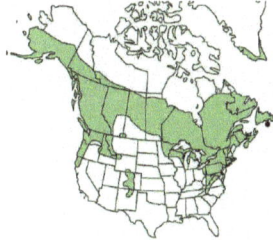

Best known of the twayblades, but so small and inconspicuous that it has been discovered only by the more persistent of its seekers, this moisture loving orchid grows from Greenland to the Pacific Northwest and to New Mexico. It has a preference for cool temperatures and is found most plentifully in the north, in the sphagnum moss of cold evergreen swamps, and, less commonly, in groves of hemlock or moist woods along Lake Superior. It blooms early, in May and June, before the full heat of summer arrives.

So well does its greenish-purple coloring combine with its low stature to make it almost invisible, that one may easily be no more than two feet away yet miss it altogether, unless, by some bit of good fortune, sunlight pierces its shadowy habitat to reveal it. Plants are not especially attractive but of interest to all lovers of orchids because of its small size and comparative rarity.

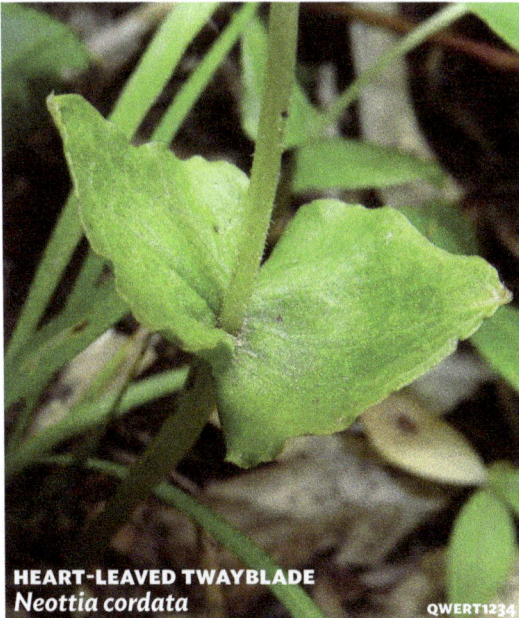

HEART-LEAVED TWAYBLADE
Neottia cordata
QWERT1234

Often growing in moist loose sphagnum, perhaps in a cool dark pocket of some remote cedar swamp, it is frequently submerged in moss nearly up to its paired roundish-oval or somewhat heart-shaped leaves. The entire plant is but eight inches high and the loose raceme of almost microscopic flowers occupies the upper fourth of it. Both petals and sepals are oval and wide-spreading. The lip, which is split into a pair of spreading prongs, bends sharply downward and bears on either side at its upper end a pair of curved erect horns.

HEART-LEAVED TWAYBLADE
Neottia cordata

CPTCV

REIN-ORCHID
Platanthera

THE REIN-ORCHIDS DERIVE THEIR NAME from the shape of their lip or their spur which resembles a rein or thong. Tall plants with fibrous or fleshy tuberous roots, the rein-orchids are readily identified as a group by their long and sometimes dense racemes of single colored flowers, each of which has a long spur.

The various species are readily recognized by the color of their flowers, which very nearly run the gamut of the spectrum: green, purple, orange, reddish-pink, and white each characterize different species of rein-orchid. Further distinctions are in the character of the lip, which in some species is cut or cleft, in others lobed, and in some deeply and intricately fringed. Most rein-orchids have leafy stems, but a few species bear their flowers on a tall naked stalk arising between basal leaves.

Fifteen species of rein-orchid are reported to grow throughout the state of Michigan (plus several hybrids). Some species prefer the zone of cool northern conifers; others confine their limits to the warmer open bogs and meadows of the lower peninsula. While each species has distinct preferences, the rein-orchids as a whole are cosmopolitan in their choices of habitats, although they are partial to wet ground. Bogs, woodlands, meadows, and swales all count certain species as their own. Nearly five hundred species of *Platanthera* grow throughout the world.

ADDITIONAL SPECIES
Alaska Rein-orchid

Platanthera unalascensis (Spreng.) Kurtz
A species of the western United States, found in eastern North America only in Chippewa and Mackinac counties and in eastern Canada (Anticosti Island, Quebec, Lake Huron shoreline region of Bruce County, Ontario). In Michigan, its typical habitat is on shallow soils over limestone, and is most commonly found on Drummond Island. Plants usually have 2–3 lance-shaped leaves at their base, these withering during or before flowering; flowers are green and very small ($^3/_{16}$ inch), and loosely spaced on the spike. Plants often have an unpleasant ammonia-like smell. This orchid was previously named *Piperia unalascensis*.

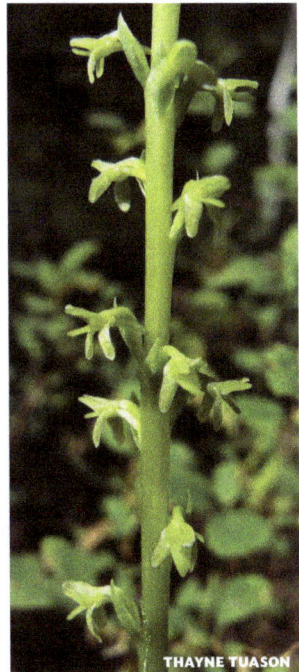

ALASKA REIN-ORCHID
Platanthera unalascensis

THAYNE TUASON

ANDREWS' ROSE-PURPLE ORCHID
Platanthera ×andrewsii (M. White) Luer

SYNONYM *Habenaria andrewsii* White

This rare orchid is a true hybrid which resulted from a cross between the **Ragged** (*Platanthera lacera*) and the **Small Purple Fringed** (*P. psycodes*) **Orchids**. It has been reported from Emmet and Keweenaw counties. In North America, it is reported from northeastern Canada, south to Wisconsin, Michigan, North Carolina, and Maine.

The fortunate few who find this species in the wet meadows which are its home will be rewarded by the sight of tall, leafy-stemmed plants bearing spikes of rose-tinted white flowers with lips lacerated like those of its ragged parent. Morris and Eames acclaim it the most famous of all rein-orchid hybrids both for its rarity and its beauty.

ANDREWS' ROSE-PURPLE ORCHID
Platanthera ×andrewsii

CHRISTINE123

TALL NORTHERN GREEN ORCHID
Platanthera aquilonis Sheviak

SYNONYM *Habenaria hyperborea* auct. non (L.) R. Br.

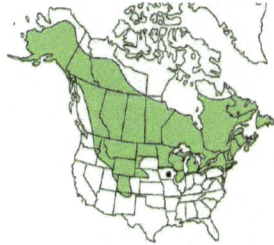

Not uncommon, yet frequently confused with the Tall White Bog Orchid, this tall leafy-stemmed northern orchid inhabits bogs and wet woods, streamsides and moist gravelly places from Labrador to Alaska. In sun or shade it blooms from June through August. Morris and Eames describe it as hardy and prolific. Certainly it is stout, for it often attains a height of two to three feet and being able to grow well in either acid or alkaline soil, its range is wide. One often finds this inconspicuous orchid beside rotted logs.

The leaves are erect, lance-like, sheathing at the base and partially clasping the stout but hollow stem. As in many *Platanthera,* the leaves decrease in size as they approach the floral spike, where a leafy bract subtends each flower.

The ovaries are twice as long as the petals and sepals, linear-cylindrical, ribbed, and slightly twisted. The sepals are darker green and thicker than the yellowish-green petals; the upper one combines with the petals to form a hood-like protection over the column. The lip is not dilated but protrudes tongue-like and is blunt at the tip. The one-fourth inch long slender spur is incurved.

The pollen masses are so arranged that they frequently fall out of the anther cells and on to the stigma so that the orchid fertilizes itself. However, it is also cross pollinated by insects.

The flowers are unscented, in contrast to the similar *Platanthera huronensis,* whose flowers have a sweet fragrance.

TALL NORTHERN GREEN ORCHID
Platanthera aquilonis

DOUG MCGRADY

WHITE FRINGED ORCHID
Platanthera blephariglottis (Willd.) Lindl.

SYNONYM *Habenaria blephariglottis* (Willd.) Hooker

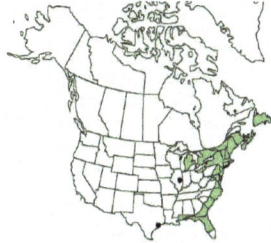

The **White Fringed Orchid** differs but slightly from the rare **Yellow Fringed Orchid** (*Platanthera ciliaris*). It is not quite as tall, the floral spike is less crowded, and the individual flowers are smaller. The lip is somewhat narrower and the fringe is finer in the White Fringed Orchid. Habits of growth, time of bloom, and soil preferences are the same for the two species. However, this orchid is more boreal in its range and the farther north one finds it, the more it is secluded in inaccessible bogs of sphagnum moss, withdrawing from the open, moist, sandy situations which it shares with the Yellow Fringed Orchid in the southern portions of its range south of Michigan.

The name 'blephariglottis' comes from two Greek words meaning literally "fringed throat," so it is therefore not accurately descriptive, since it is not the throat but the petals and lip of the flower which are fringed.

White Fringed Orchid is known to hybridize with *Platanthera ciliaris*, resulting in an intermediate form — *P. × bicolor* — having yellow flowers and reported from the southwestern Lower Peninsula.

Nature lovers have claimed the glistening waxy whiteness of this orchid the standard of whiteness for flowers and pridefully attest that beside it the snowy petals of the white water lily show yellow tones.

In Michigan, this orchid is known only from the Lower Peninsula, and is always found in wet, open, sphagnum bogs.

WHITE FRINGED ORCHID
Platanthera blephariglottis

DOUG MCGRADY

YELLOW FRINGED ORCHID
Platanthera ciliaris (L.) Lindl.

SYNONYM *Habenaria ciliaris* (L.) R. Br.
STATUS Michigan Endangered

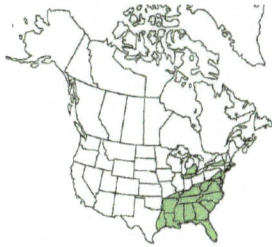

The tall leafy stems of the **Yellow Fringed Orchid** are found in July and August in sphagnum moss bogs, and rarely along the banks of streams, in open, grassy, wet meadows, and on the sunny fringe of damp thickets in sandy acid soil.

Plants are topped by three to six inch cylindrical spikes, often half as wide as long, of showy orange blossoms. Closely clustered in this often rounded spike, the flower buds resemble golden balls. When expanded, they present a feathery fringed bloom unrivalled in beauty. The upper rounded sepal bends forward, nearly concealing beneath it the two narrow petals whose toothed ends are barely visible. The paired sepals curve backward to the sides of the ovary, leaving in full view the flat, oblong, intricately-fringed lip. Here the pollinating insect alights and, swaying the long lip somewhat with its weight, finds the fringe helpful in regaining its balance. While the flowers are a ½ inch in length, their spurs are twice as long.

Two feet or more in height, the stem of the Yellow Fringed Orchid is sheathed below by long, narrow, pointed leaves which average eight inches in length for the lower ones but diminish so abruptly after the first two or three that they resemble large bracts. This semi-naked appearance of the upper stem makes the splendor of the fringed orange flower cluster even more obvious.

This orchid is found over a wide range of the United states but is restricted to the regions south of the Great Lakes and St. Lawrence River. In Michigan it is found in the lower half of the Lower Peninsula.

YELLOW FRINGED ORCHID
Platanthera ciliaris

ALLEFANT

SMALL GREEN WOOD ORCHID
Platanthera clavellata (Michx.) Luer

SYNONYMS *Gymnadeniopsis clavellata* (Michx.) Rydb., *Habenaria clavellata* (Michx.) Spreng.

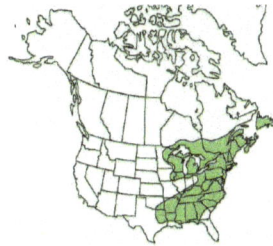

This low growing wood orchid, averaging twelve to eighteen inches in height, is found plentifully throughout much of eastern North America. Minnesota is its western outpost.

Known also as the **Little Club-Spur Orchid**, this inconspicuous, moisture-loving species has an interesting, broadly wedge-shaped lip which has three short blunt lobes at its tip. The club-shaped spur, from which the orchid derives one of its common names, is about a ½ inch long and somewhat curved.

The greenish-white flowers, rarely exceeding fifteen in number, are borne in a loose cylindrical raceme. Morris and Eames stated that the individual flowers "slewed round a quarter-turn so that the lip stands at the side." This peculiar position of the flowers tilts the nectaries to a convenient angle for insect visitors.

The Small Green Wood Orchid is a slender plant with an angled stem bearing near its base but one or two clasping, narrow blunt leaves from two to five inches long. Frequently the upper part of the stem is bare, or sometimes it has two or three linear bracts.

From mid-June throughout the summer one may find this small orchis in shady, damp woodlands and in open, sunny, mossy bogs. It seems to prefer the deep shade of the forest but can adapt itself to more open situations provided it has sufficient moisture.

Although like most orchids it is insect pollinated, the Small Green Wood Orchid has the unusual custom of maturing its pollen several days before the flowers open. By the time the petals and sepals unfold the pollen masses are loose, apparently indicating that this species is self-fertilized; nevertheless, this is not the case, for the position of the stigma is such that it cannot hold pollen which might fall from the anthers.

SMALL GREEN WOOD ORCHID
Platanthera clavellata

DOUG MCGRADY

TALL WHITE BOG ORCHID
Platanthera dilatata (Pursh) Lindl. ex L.C. Beck

SYNONYM *Habenaria dilatata* (Pursh.) Hooker

Through June and July in open cedar and tamarack bogs, masses of this tall leafy-stemmed orchid, with its long spikes of small pure white flowers is a sight one never forgets. Easily distinguishable from a distance of forty feet (even when growing in tall grass) by the whiteness of the compact cylindrical spike and its elevation above its immediate associates, it is a delightful sight to find this white orchis generously sprinkled in a bog opening. This is particularly true after tortuously picking one's way around cranberry covered sphagnum hummocks in order to avoid contact with prolific Poison Sumac, which also has a fondness for wet bogs. This is a hardy orchid tolerant of peat or marl but requiring plenty of moisture. Often to find the roots one must search through several inches of water.

In general very like the **Tall Northern Green Orchid**, the **Tall White Bog Orchid**, nevertheless, has positively distinguishing and unmistakable characters. Although these are in no wise superficial, many early botanists thought the white orchis and this green one identical, with the floral color differences due to the presence of light or shade. By its distinctly dilated lip the white orchis is known. On either side toward the base the lip is distended, while that of the Tall Northern Green Orchid tapers uniformly from base to tip.

While the pollen masses of both species fall out rather easily from their enveloping sacs, only those of the green species can fall on the stigma of the same flower and effect self-pollination. Pollen masses of the white bog orchis cannot fall on the receptive surface of its own stigma. These flowers must be cross pollinated.

A delightful, spicy fragrance redolent of cloves, or as Morris and Eames state, of syringa and cloves combined, has assured this stately and remarkably beautiful orchid a place of high regard.

TALL WHITE BOG ORCHID
Platanthera dilatata

JACOB FRANK

TUBERCLED ORCHID
Platanthera flava (L.) Lindl.

SYNONYM *Habenaria flava* (L.) R. Br.

The **Tubercled Orchid**, while similar to the **Long-Bracted Orchid** (*Dactylorhiza viridis*) in color and in habit of growth, differs in its choice of habitat. Rejecting woodlands, it prefers open wet places, favoring low, wet, mucky meadows subject to periodic flooding, and low marshy banks of running streams.

Through June and July, its tall, stout, leafy stems may be found bearing perhaps fifty minute greenish-yellow flowers in a compact spike eight inches long. The petals and sepals are similarly colored and are oval or roundish. The upper sepal and the petals form an over-arching hood, while the lower sepals expand widely on either side of the lip.

Leaves of the Tubercled Orchid are elliptic to lance-shaped; the longer lowest ones may be a foot in length. They clasp the stem and diminish in size abruptly until they become mere bristle-like bracts in the floral spike.

It is the two-lobed lip which gives this orchid its distinguishing character, for it alone of all the rein-orchids bears a protuberance on its lip. This small nose-like tubercle lies along the middle line of the lip near its base and projects upward and backward, nearly touching the column between the two anthers and at the same time extending over and dividing the opening to the slender nectar bearing spur, which exceeds the lip in length.

The tubercle is a special adaptation to insure cross pollination, for in this orchis while the anthers are parallel and almost in a line with the lip, their front ends are depressed and their sticky discs are in grooves slightly lower than the base of the tubercle. The shape and position of the discs and tubercle are so correlated that to obtain nectar the visiting insect must alight on one side or the other of the protuberance with its proboscis consequently sliding into the adjacent disc-bearing

groove. As soon as the nectar-sucking tube enters the groove, the pollen disc clasps it and is inevitably withdrawn by the insect. Obliged to make this oblique approach to the nectary, only one pollen mass can be touched by an insect at a single visit. Self-pollination cannot occur.

TUBERCLED ORCHID
Platanthera flava

ICOSAHEDRON

HOOKER'S ORCHID
Platanthera hookeri (Torr. ex A. Gray) Lindl.

SYNONYM *Habenaria hookeri* Torrey

Hooker's Orchid, one of the earliest flowering of the rein-orchids, blooms during June in cool woodlands in the northeastern part of the continent. Described as tolerating moist or dry situations, Morris and Eames concluded that this orchid distinctly prefers dry, deciduous woods rather free from underbrush where the deep leaf carpet is clear or at least only sparsely covered with low growth.

Hooker's Orchid is recognized by its pair of large, thick, shining, rounded leaves, three to five inches across, which either lie flat on the ground or rise but slightly from it. From between the leaves a gently twisted and fluted naked stalk ascends to a height of twelve to fifteen inches, the upper four to eight inches of which is a loosely flowered raceme of greenish, uniquely shaped flowers. Each flower of the cluster is large, being about two-thirds of an inch long. Although the orchid is named for an eminent English botanist, Sir William Hooker, it might well have received its very appropriate name from the structure of its flowers.

The narrow lance-shaped petals and the upper sepal bend forward over the column, while the two lower sepals are curved upward, giving the flower, in profile at least, an appearance resembling grappling hooks or widely curved tongs. The slender inch-long spur tapers uniformly to a sharp point and extends downward.

The anther cells are quite wide apart so a nectar sipping moth might drink and leave without touching them were it not for a prominent ridge extending down the middle of the stigma which virtually divides the throat of the flower in half and compels the insect to light on one side or the other. In so doing a pollen mass is inevitably detached and transferred to the next flower visited.

To those who, unfamiliar with either species, might confuse Hooker's with the very similar Large Round-leaved Orchid, Hooker's Orchid blooms fully three weeks earlier and is usually on the wane, if not entirely through blooming, when the larger species is in flower.

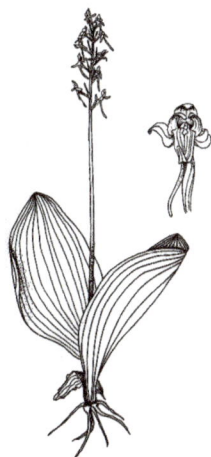

HOOKER'S ORCHID
Platanthera hookeri

NC ORCHID

LAKE HURON GREEN ORCHID
Platanthera huronensis (Nutt.) Lindl.

SYNONYM *Habenaria dilatata* var. *media* (Rydberg) Ames

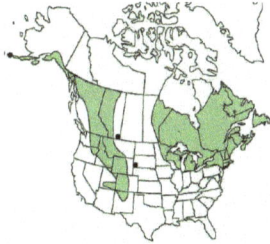

Thought to have descended from a cross between *Platanthera aquilonis* and *Platanthera dilatata*, this species shares their range of distribution and their time of bloom. It resembles both its parents, but can be easily distinguished from either of them by careful examination of its flowers, which are greenish-white and have a lip dilated at the base.

The flowers are typically intensely fragrant. It can be found in wet meadows and woodlands, marshes, fens, bogs, and along riverbanks and roadsides.

To help distinguish Lake Huron Green Orchid from the similar *Platanthera aquilonis,* note that *P. aquilonis* is essentially scentless, and the lip of *P. huronensis* is paler, often a greenish-white, while the lip of *P. aquilonis* is greenish-yellow.

JOHN GAME

JOSHUA MAYER

RAGGED FRINGED ORCHID
Platanthera lacera (Michx.) G. Don

SYNONYM *Habenaria lacera* (Michx.) Loddiges

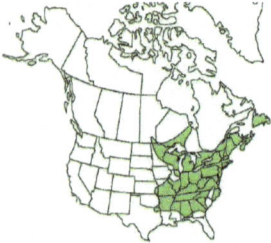

In marshy swales and meadows and in wet bogs and fens, inconspicuous among the tall grass and sedges with which it grows, thrives a tall, leafy, green orchid. Least colorful of all the fringed orchids, this one finds its chief attraction in its ragged fringed lip. So deep are its lacerations that botanists have been at their vocabularies' end to describe them. One English writer describes the lip as "elegantly jagged" in appearance, Morris and Eames refer to it as "tattered and torn," while Fuller speaks of it as "shredded."

The irregularly linear lip is really divided into three segments, each of which is wedge-shaped and further divided and subdivided into minute, almost thread-like fringes. Its lip and its greenish color render the Ragged Fringed Orchid unmistakable.

From one to two feet high, this orchid bears the lance-shaped clasping leaves, characteristic of our *Platanthera,* which are five to eight inches long near the base of the plant but decrease in size as they approach the loose spike of flowers. Individually the blooms are a ¼ to ½ inch long; from ten to fifty of them compose a cylindrical flower cluster from three to six inches in length.

Ragged Fringed Orchid blooms in July, and while not uncommon in Michigan, it does not have the sturdy, lush quality which it attains in the New England states. Morris and Eames describe their Michigan specimens as "spindly." These authors report finding the Ragged Orchid growing in a variety of wet habitats.

RAGGED FRINGED ORCHID
Platanthera lacera

DOUG MCGRADY

PRAIRIE WHITE FRINGED ORCHID
Platanthera leucophaea (Nutt.) Lindl.

SYNONYM *Habenaria leucophaea* (Nutt.) A. Gray
STATUS Michigan Endangered, federally listed as Threatened

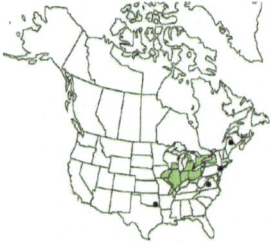

This rare orchid has been found in open fens, floating sedge mats, and wet prairie soil in Michigan's Lower Peninsula.

A tall, very fragrant orchid, it resembles somewhat both the **Ragged Fringed** and the **White Fringed Orchids** but differs enough from both of them to be readily identified.

It is a stout plant with an angled stem clasped by the lanceolate leaves characteristic of rein-orchids. The flowers are white but unlike the other white fringed species, the lip of this one is divided into three wedge-shaped segments. That of the White Fringed Orchid is narrow and undivided. The fringes and the manner in which they are subdivided resemble the Ragged orchid but the difference in color is distinctive. The 1½ inch long spur of the Prairie White Fringed Orchid is conspicuous. It blooms in July and is rare enough to make its search stimulating, its discovery an event.

JOSHUA MAYER

PRAIRIE WHITE FRINGED ORCHID
Platanthera leucophaea

JOSHUA MAYER

GREATER ROUND-LEAVED ORCHID
Platanthera macrophylla (Goldie) P.M. Br.

SYNONYMS *Habenaria macrophylla* Goldie, *Platanthera orbiculata* var. *macrophylla* (Goldie) Luer

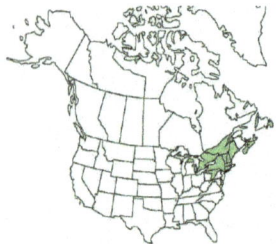

Assigned no common name by Ames, this largest of the round-leaved orchises might well be called **Greater Round-leaved Orchid**, since it is in all respects like *Platanthera orbiculata* except that it is in every way larger. By some botanists this orchis was judged merely a large form of *Platanthera orbiculata,* with which it was long combined. In blooming season, habitat, and range, the two species coincide.

ROB ROUTLEDGE

ROB ROUTLEDGE

GREATER ROUND-LEAVED ORCHID
Platanthera macrophylla

ROB ROUTLEDGE

SMALL NORTHERN BOG ORCHID
Platanthera obtusata (Banks ex Pursh) Lindl.

SYNONYM *Habenaria obtusata* (Banks ex Pursh.) Richardson

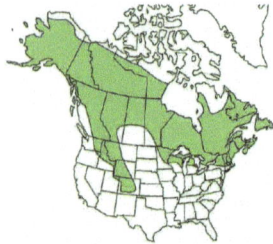

Smallest of all the rein-orchids, the little green and white flowered northern bog orchis is less than a foot in height. It is most often found in cold evergreen forests and bogs and prefers deep shade. From the middle of June through July and August this low-growing uncommon orchid may be found associated with twinflower, heart-leaved twayblades, and coralroot.

Its distinguishing feature is its single basal leaf, two to five inches long, which is blunt and broader at the tip than at the base, where it narrows into a stalk enfolding the stem. Because of the shape of the solitary leaf, this species is often called the Blunt-leaf Orchid.

The loosely-flowered two-inch spike of three to fifteen flowers, each a scarce ½ inch in length, tops a slender four-angled stem. The upper sepal is round and broad and arches over the column, while the lower ones which are not so broad are curled back. Narrow petals curve upward, the tapering slender spur curves downward.

GERTJAN VAN NOORD

The anther sacs are widely separated, and on the lip below them Gibson observed two ridges which as they approach the nectary converge into one, making it necessary for the alighting insect to swerve its tongue and turn its head, so that it inevitably leaves the flower with one pollen mass or the other.

SMALL NORTHERN BOG ORCHID
Platanthera obtusata
CHLOE AND TREVOR VAN LOON

LESSER ROUND-LEAVED ORCHID
Platanthera orbiculata (Pursh) Lindl.

SYNONYM *Habenaria orbiculata* (Pursh.) Torrey

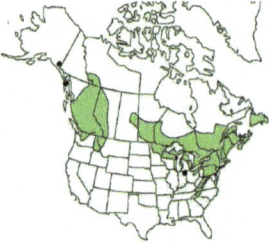

In forests of hemlock, birch, sugar maple, and balsam, and in deep mossy cedar bogs, the **Lesser Round-leaved Orchid**, which blooms during July and August, finds its preferred home. Resembling Hooker's Orchid, it can, however, be easily distinguished not only by its greater size and a somewhat later blooming season, but by evident differences in the flowers.

Two thick, flat, round leaves, often nine inches in diameter, shining and dark green above but silvery below, lie flat to the ground. Between them rises a bracted flower stem two feet high bearing at its summit a loose open spike, five to ten inches long and an 1½ to 2 inches in diameter, of large, individually long-stalked, pale greenish-white flowers.

Whereas the floral parts of Hooker's Orchid have the appearance of a claw, those of this species assume a totally different attitude. The petals and sepals are wider and the spur is blunt, almost club-shaped, and tips upward at its apex. The blunt lip of the flower, on the contrary, points downward. This is just the reverse of the positions of these parts in Hooker's Orchid, where the lip curves upward and the awl-like spur points downward. This difference in the relative position of spur and lip gives the flowers a totally changed attitude on their stems. While in both species the upper sepal bends forward over the column, only in Hooker's do the petals also arch. Those of the Lesser Round-leaved Orchid are wide spreading and curve upward.

Pollinated at night by one of the sphinx moths, this orchid holds its nectar deep within a spur fully an 1½ inch long. To obtain this, the visiting insect

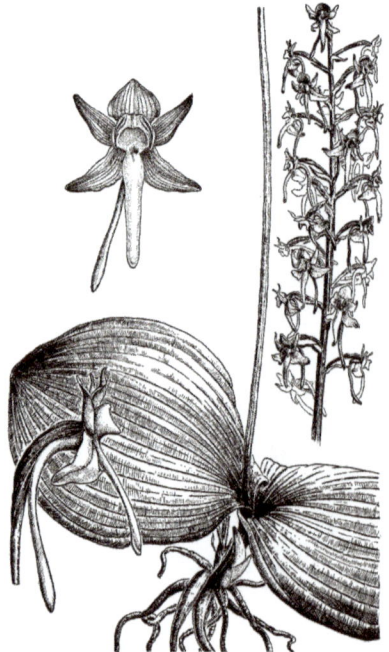

must alight on one side of the flower because the ridged stigma prevents a more central landing which would allow access to the nectary without contact with either of the widely separated pollen masses. Forced to make a lateral approach to the nectary, the insect inevitably carries a pollen mass away on its head.

LESSER ROUND-LEAVED ORCHID
Platanthera orbiculata

ALLEFANT

SMALL PURPLE FRINGED ORCHID
Platanthera psycodes (L.) Lindl.

SYNONYM *Habenaria psycodes* (L.) Sprengel

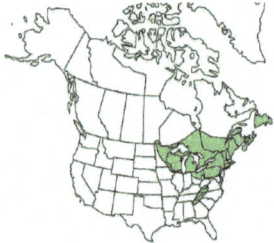

This orchid is widely distributed in wet meadows, depressions in deciduous woods, along slow streams in damp woods, and on rock ledges near Lake Superior; soils are medium acid. However, it has become rare in southernmost Michigan.

This tall, fragrant, delicately-fringed, magenta-purple orchis is considered by many to be the most showy of the rein-orchids. It is unquestionably better known than the majority of them, for to find it one need not traverse sphagnum bogs nor push through mazes of fallen cedar, as he must commonly do if the rarer of our native orchids are to be seen in their natural settings.

This stately purple orchid blooms from early July until mid-August. It grows from one to three feet in height and has the typical narrow, lance-shaped, clasping leaves common to the entire group. In this species, they are rather stout and have a prominent and heavy mid-rib. The stem is square in section and is mottled with purple. At its tip is the full-flowered raceme, averaging four inches in length. It is slender, however, like the entire plant, and is composed of as many as eighty flowers, each less than ⅓ inch in length. The outstanding character of this species is its abundantly fringed lip which is divided into three wedge-shaped segments.

MALCOLM MANNERS

SMALL PURPLE FRINGED ORCHID
Platanthera psycodes

DOUG MCGRADY

POGONIA
Pogonia

THE ORCHIDS IN THIS GROUP are low growing herbs scarcely a foot high, needing very acid situations, and represented in North America by only one of the reported five species which grow throughout the world (North America, Japan, China, and Indonesia). Their name, like that of so many plants, is derived from the Greek, and means "beard." It is appropriately descriptive, for the bearded lip is one of the Pogonias' most distinctive characters. They are further differentiated by their solitary terminal showy flowers with entirely separate petals and sepals and a club-shaped column which is free from the lip; they have no spur. One or two narrow leaves alternate on the stem. The rootstocks are slender.

ROSE POGONIA
Pogonia ophioglossoides (L.) Ker-Gawl.

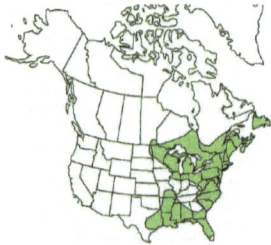

Rose Pogonia is more common than **Arethusa**, with usually paler pink flowers, and flowering a little later. Both species are found in similar boggy habitats.

Rose Pogonia is one of the most delicate and beautiful of our wild orchids. From late June until mid-August this small, sweet-scented, rose-colored orchid grows in profusion in sunny openings of grassy bogs which quite often are quaking mats of sphagnum. In wet meadows and marshes, along the grassy shores of quiet lakes in acid soil, the Rose Pogonia thrives. Although it is a low growing orchid, its perfume, which Barker declares to be like raspberries, and its color, proclaim its presence in the ferns and mosses among which it grows.

It is a nodding flower, with yellow-tipped, magenta, fringed hairs providing a raised, three-ranked crest on the upper surface of its lip. The glistening petals and sepals are alike in color and similar in form. They are elliptic to oval, the petals being somewhat broader than the sepals. There is no spur. When fully expanded the flower bears some resemblance to the open mouth of a reptile, hence its other common name, Snake-mouth.

This orchid has a most ingenious device for assuring cross-pollination. The pollen masses are enclosed in small, pouch-like recesses protected by a lid which fits like the hinged cover of a box. When the visiting insect alights on the crested lip, it pushes this lid more tightly shut as it forces its head into the low, narrow opening to the nectary. By this same act the insect's head is forced against the sticky stigma, which receives the pollen that is brought with it into the flower. With no alternate exit, the insect is forced to back out of the flower and in so doing trips open the anther lid, releasing pollen masses which cling to its head and are carried to the next flower visited. As soon as the insect's head has been dusted with pollen, the anther lid springs shut again to hold any pollen which may be left for the next accommodating visitor.

The slender stem of Rose Pogonia is rarely more than a foot in height. It is clasped midway to its summit by a single elliptic-oval leaf. Its fibrous fleshy roots require continued moisture. Where meadows and bogs are drained this orchid is one of the first to disappear.

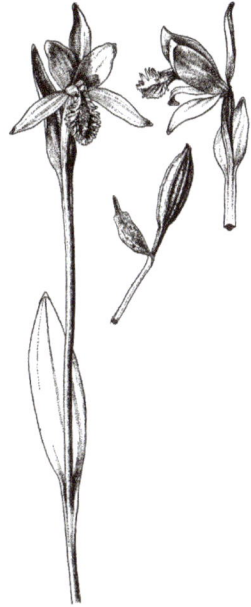

ROSE POGONIA
Pogonia ophioglossoides

ALLEFANT

LADIES'-TRESSES
Spiranthes

ABOUT 27 SPECIES OF THESE SLENDER ORCHIDS are found in North America north of Mexico; the majority of them occur in the eastern part of the United States, and ten grow in Michigan. Unlike many orchids, ladies'-tresses can thrive in open meadows. For this reason they can become established and survive after the clearing of the forest when forest species, which need shade, disappear with the destruction of the canopy.

Their scientific name, referring to the spiral flowers or to the twisted stalk, is well chosen, though it is difficult to imagine the derivation of the common name. The last orchids to bloom in the fall and among the earliest to flower in the spring, ladies'-tresses, although easy to recognize as a group, can be a challenge to identify at the species level.

In common are stiff, upright, slightly-twisted stems, bearing rather long dense whorled spikes of small white flowers (these sometimes fragrant), rise from tuberous or fleshy fibrous roots. The flowers have no spurs, and the tongue-shaped lip has a tubercle or knob on each side of its base.

Bees transfer the pollen and begin at the bottom of the spiral flower cluster and work up. This habit is an important one, for the lower flowers of the spike are the more mature.

Darwin discovered the curious mechanism by which ladies'-tresses assure cross-fertilization. When the entering bee alights on the flower's lip, it cannot obtain nectar without touching the column which is closely adjacent to it. Its slightest touch causes the beak to split, the opening to widen, and a boat-shaped anther-bearing disc to emerge and fasten itself to the insect's head. As the flowers grow older the column rises, producing a wider aperture to the nectary, which in young flowers is difficult to approach.

Coincident with inaccessible nectar is the immaturity of the stigma, which does not become sticky and receptive until the flower's throat is freely open to admit bees. In visiting first the lowest blooms of the spike, the bee, therefore, enters mature flowers whose stigmas can receive and retain the pollen it has brought. As it progresses upward, the insect receives pollen from newly opened flowers, whose own stigmas are not yet receptive, and carries it to the mature lower blossoms of the next plant visited.

APPALACHIAN LADIES'- TRESSES
Spiranthes arcisepala M.C. Pace

A newly defined species (see Pace and Cameron, 2017), believed to range throughout the northeast, from Ontario south to North Carolina and as far west as Ohio and extreme southern Michigan. Plants have one to four upright basal leaves which wither soon after flowering. The flowers are white, faintly

scented, and arranged in a loose spiral on the spike. It grows in moist to wet places, such as low prairies, bogs, marshes, and fens. Its downward arching lateral sepals help distinguish this orchid from *Spiranthes cernua* and *S. ochroleuca*.

CASE'S LADIES'-TRESSES
Spiranthes casei Catling & Cruise

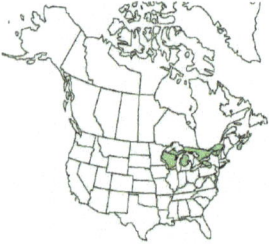

Named for Frederick W. Case, Jr. (1927-2011), student of Great Lakes' orchids and trilliums, **Case's Ladies' Tresses** is known from eastern Canada and the United States, from Wisconsin to Nova Scotia.

Plants have two to five basal and lower stem leaves (which are withered by flowering time), and its upper stem and flowers are covered in thin hairs. The spiraled inflorescence supports up to 40 white or greenish cream-colored flowers, with a leafy green bract extending out from each flower. The lip has two prominent thickened 'bumps'.

It can be found in fields, barrens, and open woodlands on sandy, acid soils, often with bracken fern and wild oat grass (*Danthonia spicata*).

Distinguished from *Spiranthes magnicamporum* by its scentless flowers and appressed, instead of spreading, sepals; and separated from *S. incurva* by its single, instead of multi-ranked, inflorescence and by its smaller flowers. Compared to *Spiranthes ochroleuca*, Case's Ladies'-Tresses is more northerly in its distribution, and blooms earlier, from early August to early September.

AARON CARLSON

AARON CARLSON

SPHINX LADIES'-TRESSES
Spiranthes incurva (L.) Rich.

SYNONYM *Spiranthes cernua* (L.) Rich.

Long treated as *Spiranthes cernua,* recent (2017) research by Pace and Cameron has determined that this new species, **Sphinx Ladies'-Tresses,** replaces *S. cernua* in Michigan and the northern and western portions of the latter's former range (*Spiranthes cernua* is now considered a species found on the east coast and in the southeastern United States).

The Sphinx Ladies'-Tresses grows to about fifteen inches in height. Its white flowers are slightly fragrant. The lip is strongly downwardly bent at about $\frac{1}{3}$ to $\frac{1}{2}$ the distance from its tip, and the hood is slightly to strongly upturned near its apex. The basal linear grass-like leaves typically wither before the flowers open.

Plants grow in a range of wet to sometimes dry habitats: bogs, fens, marshes, low wet meadows and lakeshores, frequently in company with gentians, for this is a fall-blooming species. In Michigan, it is in flower from August through October.

NATE MARTINEAU

SPHINX LADIES'-TRESSES
Spiranthes incurva

SONNIA HILL

SLENDER LADIES'-TRESSES
Spiranthes lacera (Raf.) Raf.

SYNONYM *Spiranthes gracilis* (Bigel.) Beck

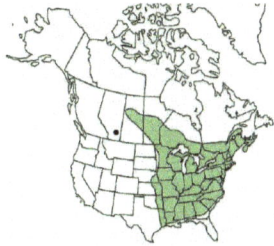

The **Slender Ladies'-Tresses** blooms from July through September in moderately acid soil of open upland woods, sandy plains and shores. A favored habitat is under jack pine with blueberries and bracken ferns.

It is a very slim, smooth species, from eight to thirty inches tall, bearing a basal cluster of oval leaves that wither before the flowers open. The tiny, numerous flowers grow in a slender twisted spike; very close together, they wind in a single row many times around the stem. The individual blooms, less than a ¼ inch long, have their petals and upper sepal hooded over the column. Lower sepals are spreading and somewhat longer than the wavy margined lip, which bears a green stripe on either side of a central groove.

In late summer and fall, *Spiranthes lacera* (and *S. tuberosa*) produce an overwintering rosette of small ovate green leaves that lie nearly flat on the ground; the leaves resemble those of the Rattlesnake Plantains, but lack their characteristic net-like pattern.

The distinctive green spot on the lip of the **Slender Ladies'-Tresses** flowers separates it from all other Michigan *Spiranthes*.

FRITZ REYNOLDS

SLENDER LADIES'-TRESSES
Spiranthes lacera

MASON BROCK

WIDE-LEAVED LADIES'-TRESSES
Spiranthes lucida (H.H. Eaton) Ames

Sometimes called **Shining Ladies'-Tresses** because of its glossy leaves, this low-growing, smooth, fleshy species prefers the wet soil of limestone regions, and is very local in its distribution. It may be found in flower in June and July (our earliest flowering *Spiranthes*) in wet meadows and along riverbanks and lakeshores. It is frequently found in dense colonies, but being less than half the height of the Slender Ladies'-Tresses, it is quite naturally less obvious.

The floral spike, though but one to three inches long, is, like the other species of ladies' tresses, crowded and twisted, and the individual flowers lose nothing in beauty because of their very small size. They are distinguished by the yellow to yellow-orange stripe edged with green which adorns the center of their flared white lip.

The smooth, fleshy, shining leaves are oblong and basal. They do not wither like those of the Sphinx and Slender Ladies'-Tresses but remain fresh throughout the season. The roots are fleshy.

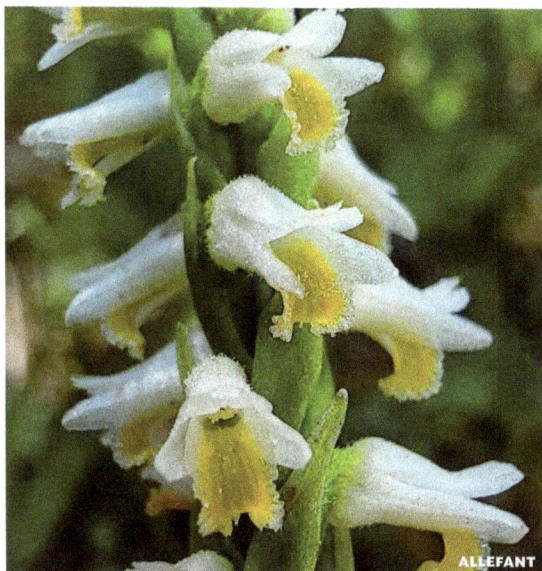

PRAIRIE LADIES'-TRESSES
Spiranthes magnicamporum Sheviak

Formerly included in the **Nodding Ladies'-Tresses** (*Spiranthes cernua*) complex of species, **Prairie Ladies'-Tresses** is the latest species of the group to flower, starting to bloom after *Spiranthes incurva* is past its prime, and often blooming well into October. It has a strong vanilla-licorice fragrance; and is found in the calcareous soils of wet meadows, fens, and moist to dryish prairies.

Also called **Great Plains Ladies' Tresses**, this *Spiranthes* is widely distributed in North America from Ontario to Mexico. Plants have two or three basal leaves which usually wither by the time of flowering in the late summer or fall. The flowers, arranged in a tight spiral, number up to 40, and are white, cream-colored, or yellowish. The lip curves sharply downwards; the center of the lip is often yellow. The strongly scented flowers help distinguish this orchid from *Spiranthes incurva*.

JOSHUA MAYER

JOSHUA MAYER

PRAIRIE LADIES'-TRESSES
Spiranthes magnicamporum

ERIC HUNT

YELLOW LADIES'-TRESSES
Spiranthes ochroleuca (Rydb.) Rydb.

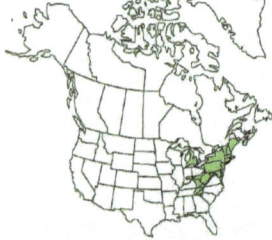

Another species formerly grouped with *Spiranthes cernua*, **Yellow Ladies' Tresses** ranges from northeastern Canada southward to South Carolina and Tennessee. It prefers the moist to dryish sandy, acid soils of open woodlands, fields, roadsides, and thickets. In Michigan it flowers in September.

Plants have three to six basal and lower stem leaves which usually persist through flowering. Flowers, up to 60 in number and spirally arranged, are small, white, yellowish, or greenish-white. The lip has finely scalloped margins and is often yellow-colored at its center.

ALLEFANT

ORCHI

ALLEFANT

OCTOBER LADIES'-TRESSES
Spiranthes ovalis Lindl.

STATUS Michigan Threatened

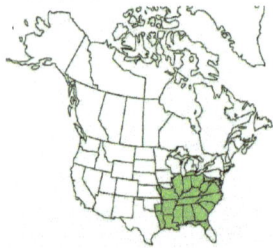

Also called called **Lesser Ladies'-Tresses** due to its small size, *Spiranthes ovalis* is widely distributed across the central and eastern United States; however, it barely extends into southern Michigan, the northernmost portion of its range.

Plants are small, growing only 5 to 14 inches tall, and have two to five basal or lower stem leaves, these usually persisting through flowering. The flowers are in a tightly spiraled inflorescence of up to 50 small, white flowers. The margins of the lip are slightly wavy.

It grows in moist fields, moist to dry thickets, open woods, fields and prairies; soils are often sandy.

The flowering season is very late, beginning in late September and extending even to early November.

October Ladies'-Tresses can be distinguished from *Spiranthes lacera* by the absence of a central green or yellow spot on the lip.

JOSHUA MAYER

OCTOBER LADIES'-TRESSES
Spiranthes ovalis

ALLEFANT

HOODED LADIES'-TRESSES
Spiranthes romanzoffiana Cham.

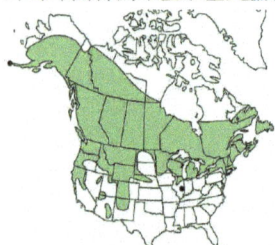

The **Hooded Ladies'-Tresses** is a distinctly northern species and transcontinental in its range. From July through September in wooded and open bogs, in moist grassy depressions, along gravelly lake shores, and in wet marl, may be found this most widely distributed of all the Ladies'-Tresses.

A shallowly rooted species, its cluster of tuberous finger-like roots are from 1 to 1½ inches long. The stem, averaging a foot but occasionally reaching fifteen inches in height, bears smooth, linear, thickish, grass-like clasping leaves which diminish in size as they approach the floral spike, until the upper ones are reduced to mere acutely tipped leafy bracts. The leaves are broader toward their outer end than at the base and have rather blunt tips.

The flowers, rarely exceeding ⅓ inch in length, are compactly arranged in three spiral ranks. Their petals and sepals are all united into a hood, while the lip is fiddle-shaped and expanded at the outer end. This fiddle-shaped lip helps distinguish Hooded Ladies'-Tresses from all other Michigan *Spiranthes*.

With the fragrance of delicate lilacs, the flowers, including their leafy bracts, have a pristine glistening quality so characteristic of certain members of the orchid family. They are so delicately thin that their cellular structure can be seen easily with the aid of a hand lens.

Sometimes called **Romanzoff's Ladies'-Tresses**, this species was named in honor of Russian statesman, Nikolay Rumyantzev, a patron of science.

ANDREY ZHARKIKH

HOODED LADIES'-TRESSES
Spiranthes romanzoffiana

ANDREY ZHARKIKH

LITTLE LADIES'-TRESSES
Spiranthes tuberosa Raf.

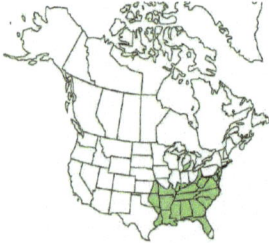

Little Ladies'-Tresses is widely distributed in the southern and central United States, from Texas to Michigan and east to Florida and Massachusetts.

Plants have three to five basal leaves which wither by the time of flowering. The inflorescence bears up to 30 small, white flowers arranged in a loose, single spiral.

It can be found in dry to moist, sandy prairies and meadows, and woodland openings, sometimes under bracken, and often in mats of haircap moss (*Polytrichum*).

This tiniest *Spiranthes* resembles a miniature *S. lacera*, but with a pure white lip. It generally blooms later than *Spiranthes lacera,* and can be distinguished from other similar *Spiranthes* species by its pure white lip.

In late summer and fall, Little Ladies'-Tress produces an overwintering rosette of small green leaves that lie nearly flat on the ground; the leaves resemble those of the Rattlesnake Plantains, but lack their characteristic net-like pattern.

SONNIA HILL

LITTLE LADIES'-TRESSES
Spiranthes tuberosa

ERIC HUNT

CRANEFLY ORCHID
Tipularia

TIPULARIA IS A GENUS of terrestrial orchids containing three species of which only one, *Tipularia discolor,* is found in North America; the other two are found in the Himalayas and Japan. In North America, *Tipularia* is distributed throughout the eastern and central United States, from Florida to Massachusetts and as far west as eastern Oklahoma and eastern Texas. It produces a leaf in the fall which stays green throughout the winter; the leaf then senesces in the spring before the plant flowers, leafless, in the summer. The leaf upper surface is green with dark purple spots and the underside has a distinctive dark purple pigmentation. It produces up to fifty-five small, somewhat asymmetrical flowers that are generally greenish yellow and sometimes tinged with purple.

CRANEFLY ORCHID
Tipularia discolor (Pursh) Nutt.

STATUS Michigan Endangered

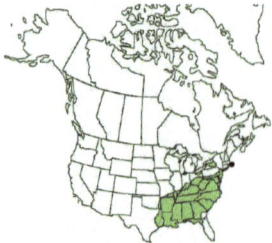

Cranefly Orchid produces a single leaf in the fall which stays green throughout the winter; the leaf upper surface is purple-spotted and the underside is pigmented with purple. The leaf wilts in the spring before the plant flowers in the summer. The spotted leaves and nectar spurs on the flowers help distinguish this orchid from **Puttyroot** (*Aplectrum hyemale*), which also has a winter-green leaf.

In Michigan it grows in rich, sandy deciduous woods of beech-sugar maple or oak, mostly just inland from Lake Michigan dunes.

A very distinctive orchid, the winter-green leaf (as in *Aplectrum*) withers in spring before the flowers mature in the summer. Often, an old fruiting stalk may be found with a well-developed leaf at the base. Underground, plants have a rhizome with a series of tubers.

Michigan is at the northern edge of this species' range, and while recorded from Berrien County in southwestern Michigan, this plant has apparently not been observed there since 2005.

ALPHAWOLF

DANIELLE ...

ERIC HUNT

CRANEFLY ORCHID
Tipularia discolor

JUDY GALLAGHER

NODDING POGONIA
Triphora

APPROXIMATELY 18 SPECIES OF TRIPHORA have been described from Central and South America. In North America, one species is found in the eastern United States and southern Ontario. It was formerly classified as a *Pogonia*, but because the flowers are technically different, this group has been given a separate generic name. These plants are low, few-flowered herbs with fleshy tubers.

NODDING POGONIA
Triphora trianthophoros (Sw.) Rydb.

STATUS Michigan Threatened

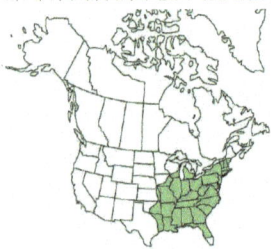

This orchid is also known by the common names **Three-birds** or **Nodding-caps**. Its specific name *trianthophora*, from the Greek, refers to the three flowers it usually bears simultaneously. A fading flower below, one in its prime of full bloom above, and an unopened bud at the tip, comprise the inflorescence.

Nodding Pogonia is a small, slender, delicate orchid, eight inches high, whose stem is entirely clasped by two to eight very small, cupped, oval leaves about ½ inch across. These vary in habit according to the season. Fuller states that "during a moist season they are fully developed, while in a dry one green coloring is absent from the leaves which become scale-like and the entire plant takes on a purplish-pink color." It appears almost saprophytic.

The flowers on slender stalks are borne in the axils of the leaves. They are about ½ inch long and are pale pink to white or slightly magenta-tinged. Both petals and sepals are lance-shaped and spreading. The oval, somewhat three-lobed lip is not crested like that of Rose Pogonia but bears on its upper surface three prominent bright green nerves. After pollination the edges of the lip curve inward. The deep rich magenta-violet pollen is fixed and not released by a hinged anther lid as in the Rose Pogonia.

Preferring deep, humus-filled hollows in mature forests of beech, maple, or oak, the Nodding Pogonia blooms in August and September.

Found across much of the eastern United States, this orchid is nevertheless rare and has been recorded from only a limited number of counties in the

Lower Peninsula. Its appearance from year-to-year is sporadic, sometimes plentiful in a small area and non-existent a short distance away. Furthermore, where some botanists have reported it growing in abundance, others could not find a single plant.

Lownes explains the peculiar distribution and intermittent appearance of this rare orchid, of which he saw upwards of twenty thousand plants within the area of a square mile. In the same place the following year there were but five hundred; the next year there were only two plants; the succeeding year none. Two years later the plants were as abundant as the first year he saw them.

This, Lownes believes, is due to the root system, which he states is different from that of any other orchid found in northern United States: "It occurs in the upper strata of humus apparently never penetrating to the soil below. Essentially it consists of a single ovoid translucent white tuber. From the lower portion of the flowering stem, a number of stolons grow out in a horizontal direction, each bearing at the end a smaller tuber. A well-grown plant may have ten or twelve of these stolons, with tubers ranging in size from a pinhead to a third of an inch or more in length. The whole plant is very brittle, and this is true especially of the root system and the stolons. If, however, a plant is carefully removed from the humus in which it grows, and the root system studied, a small scar will be found at the tip of the mature tuber. Each plant apparently blooms but once. The subsequent decay of the mature tuber cuts off the small tubers at the ends of the stolons; each of these in due season produces a bud, which in turn forms a new plant. In this way the large colonies are formed and thus, too, the extraordinary periodicity is explained."

Further peculiarities are described by Lownes: "No sign of the species can be seen until the end of July or early August. Then the stiff sharp tip of the stem, with the leaves folded closely about it, pushes its way through the leaf mold. In about a week or ten days the blossoms appear. At night and during unfavorable weather conditions the blossoms close and nod. When conditions are right, every plant opens its flowers wide, and the flower stalk is erect; this occurs, however, but once or twice during the season. Fertilization is accomplished by small bees, but the seed is rarely if ever ripened in the northern states."

JIM FOWLER

NODDING POGONIA
Triphora trianthophoros

ALLEFANT

KEY TO MICHIGAN ORCHIDS

To enable you to more accurately identify unknown orchids, use this key to narrow the possibilities to only a single species. At each pair of numbered statements (couplets), a choice is made; read both statements and pick the most appropriate for your plant of interest. Continue in this manner until the species name is given. Verify your identification by using the text descriptions, photographs, and range maps. The key includes four of the most common **Ladies'-Tresses** (*Spiranthes*) found in Michigan. See page 142 for a complete, but more technical key to Michigan's ten species of *Spiranthes*.

1   Plants with little or no green color 2
1   Plants with green color 5

2   Flowers yellowish-green, in racemes; statewide .. .. **EARLY CORALROOT** .... (*Corallorhiza trifida*)
2   Flowers striped or spotted with purple, in racemes 3

EARLY CORALROOT

3   Petals and sepals striped with madder-purple; northern Michigan .... **STRIPED CORALROOT** .... (*Corallorhiza striata*)
3   Petals and sepals marked with purple, but not striped 4

4   Lip three-lobed; statewide .. **LARGE CORALROOT** .... (*Corallorhiza maculata*)
4   Lip entire; mainly Lower Peninsula .... **SMALL CORALROOT** .... (*Corallorhiza odontorhiza*)

STRIPED CORALROOT

5   Plants with leafy stems 6
5   Plants with basal leaves only 32

6   Plants with one stem leaf 7
6   Plants with two to many stem leaves 9

LARGE CORALROOT

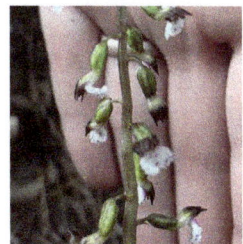

SMALL CORALROOT

7 Leaf oval, near middle of stem; flower solitary, magenta-pink; statewide . . . . **ROSE POGONIA**
. . . . (*Pogonia ophioglossoides*)

7 Leaf oval, near middle of stem; flowers in raceme, green or greenish-white **8**

8 Leaf oval, near middle of stem, flowers in raceme, green; northern Michigan . . . .
. . . . **GREEN ADDER'S-MOUTH**
. . . . (*Malaxis unifolia*)

ROSE POGONIA

GREEN ADDER'S-MOUTH

8 Leaf oval, near base of stem, flowers in raceme, greenish-white; statewide . . . .
. . . . **WHITE ADDER'S-MOUTH**
. . . . (*Malaxis monophyllos*)

9 Plants with two stem leaves **10**
9 Plants with many stem leaves **13**

10 Leaves alternate; statewide . . . .
. . . . **SMALL GREEN WOOD ORCHID**
. . . . (*Platanthera clavellata*)

10 Leaves opposite (or slightly alternate in *Neottia convallarioides*) **11**

WHITE ADDER'S-MOUTH

11 Leaves oval-elliptical; flowers pale whitish-green; Upper Peninsula . . . . **AURICLED TWAYBLADE**
. . . . (*Neottia auriculata*)

11 Leaves broader; flowers purple-tinged or yellowish-green; Upper Peninsula and northern Lower Peninsula **12**

SMALL GREEN WOOD ORCHID
Platanthera clavellata
ROB ROUTLEDGE

SMALL GREEN WOOD ORCHID

AURICLED TWAYBLADE

12 Leaves roundish-oval, somewhat heart-shaped, flowers watery-purple; northern Michigan . . . . **HEART-LEAVED TWAYBLADE** . . . . (*Neottia cordata*)

12 Leaves opposite , roundish-oval, flowers yellowish-green; northern Michigan . . . . **BROAD-LEAVED TWAYBLADE** . . . . (*Neottia convallarioides*)

HEART-LEAVED TWAYBLADE

13 Leaves whorled; Lower Peninsula . . . . **WHORLED POGONIA** . . . . (*Isotria verticillata*)

13 Leaves alternate **14**

14 Leaves wide-oval, concave; Lower Peninsula . . . . **NODDING POGONIA** . . . . (*Triphora trianthophoros*)

14 Leaves linear-lance shaped. **15**

BROAD-LEAVED TWAYBLADE

15 Leaves only near base of stem, usually withering when flowers mature; statewide. . . . . **HOODED LADIES'-TRESSES** . . . . (*Spiranthes romanzoffiana*)

15 Leaves along entire stem **16**

WHORLED POGONIA

16 Leaves conspicuously parallel-ribbed, somewhat roughish, flowers with a pouch. **17**

16 Leaves not conspicuously parallel-ribbed, smooth, often glistening **21**

17 Flowers yellow **18**

17 Flowers white and white with rose-pink and white **19**

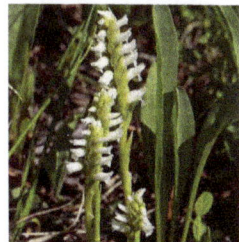
HOODED LADIES'-TRESSES

18 Lip light yellow, ¾ to 1¼ inches long, small plant 8–24 inches tall, fairly smooth, fragrant; mostly Lower Peninsula. . . . . **SMALLER YELLOW LADY'S-SLIPPER** . . . . (*Cypripedium parviflorum* var. *makasin*)

SMALLER YELLOW
LADY'S-SLIPPER

18 Lip golden yellow 1½ to 2 inches long, tall plant 9–28 inches high, conspicuously downy; statewide . . . . **LARGER YELLOW LADY'S-SLIPPER** . . . . (*Cypripedium parviflorum* var. *pubescens*)

LARGER YELLOW LADY'S-SLIPPER

19 Lip pure white, smooth, purple-veined within, ⅝ to 1 inch long, low plant 6–12 inches high; southern Lower Peninsula . . . . **SMALL WHITE LADY'S-SLIPPER** . . . . (*Cypripedium candidum*)

19 Lip white and also shaded rose-pink, or lip white with crimson veins **20**

SMALL WHITE LADY'S-SLIPPER

20 Lip white, crimson-veined, prolonged at apex into blunt spur, mouth of lip covered with white woolly hair; Upper Peninsula and northern Lower Peninsula . . . . **RAM'S HEAD LADY'S-SLIPPER** . . . . (*Cypripedium arietinum*)

20 Lip white to pale purple, rose-pink in front, 1½ inches long, tall plant 1–3 feet high; statewide. . . . . **SHOWY LADY'S-SLIPPER** . . . . (*Cypripedium reginae*)

RAM'S HEAD LADY'S-SLIPPER

21 Lip of flower fringed **22**
21 Lip of flower not fringed **27**

22 Flowers yellow-orange; southern Lower Peninsula . . . . **YELLOW FRINGED ORCHID** . . . . (*Platanthera ciliaris*)

22 Flowers white, greenish-white, or purple **23**

SHOWY LADY'S-SLIPPER

YELLOW FRINGED ORCHID
Platanthera ciliaris

NC ORCHID

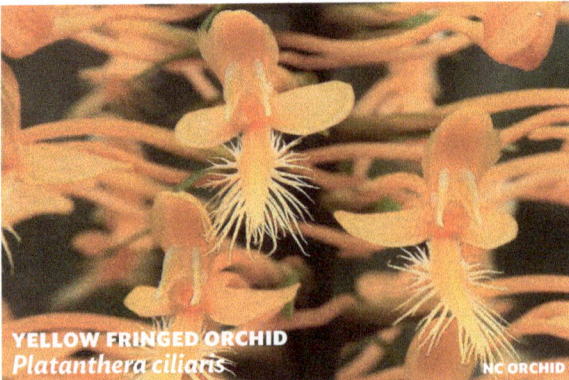

YELLOW FRINGED ORCHID

23   Flowers purple; statewide . . . .
. . . . **SMALL PURPLE FRINGED ORCHID**
. . . . (*Platanthera psycodes*)

23   Flowers white, greenish-white, or white tinted with
rose **24**

SMALL PURPLE
FRINGED ORCHID

24   Flowers pure white **25**
24   Flowers not pure white, either greenish-white or
white tinted with rose. **26**

25   Lip 3-parted; rare in southern Lower Peninsula . .
. . . . **PRAIRIE WHITE FRINGED ORCHID**
. . . . (*Platanthera leucophaea*)

25   Lip entire; Lower Peninsula . . . .
. . . . **WHITE FRINGED ORCHID**
. . . . (*Platanthera blephariglottis*)

PRAIRIE WHITE
FRINGED ORCHID

26   Flowers white, tinted with rose; uncommon, re-
ported from Emmet and Keweenaw counties . . . .
. . . . **ANDREWS' ROSE-PURPLE ORCHID**
. . . . (*Platanthera x andrewsii*)

26   Flowers greenish-white, very finely cut; statewide
. . . . **RAGGED FRINGED ORCHID**
. . . . (*Platanthera lacera*)

WHITE FRINGED ORCHID

ANDREWS' ROSE-PURPLE
ORCHID

RAGGED FRINGED ORCHID
Platanthera lacera

USGS

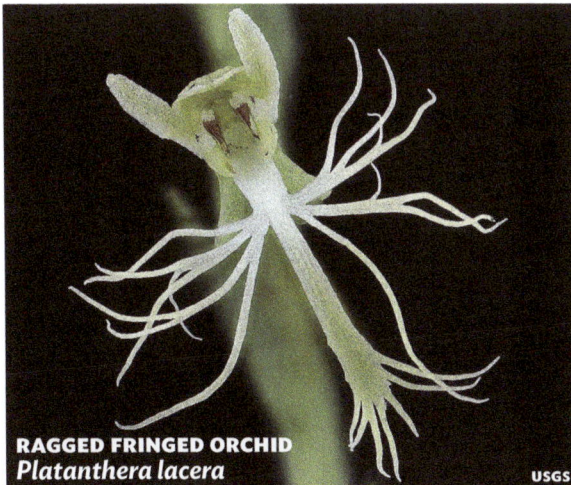

RAGGED FRINGED
ORCHID

27   Flowers pure white, fragrant; statewide . . . .
. . . . **TALL WHITE BOG ORCHID**
. . . . (*Platanthera dilatata*)
27   Flowers greenish **28**

28   Lip entire **29**
28   Lip lobed **30**

29   Lip of flower constricted at middle; statewide (our
only introduced orchid) . . . . **HELLEBORINE**
. . . . (*Epipactis helleborine*)
29   Lip of flower not constricted at middle **31**

30   Lip widened and rounded at base; mostly Upper
Peninsula and northern Lower Peninsula . . . .
. . . . **LAKE HURON GREEN ORCHID**
. . . . (*Platanthera huronensis*)
30   Lip tapering gradually to a point; statewide . . . . . .
. . . . **TALL NORTHERN GREEN ORCHID**
. . . . (*Platanthera aquilonis*)

31   Lip 2-lobed, with a tubercle near its base; Lower
Peninsula . . . . **TUBERCLED ORCHID**
. . . . (*Platanthera flava*)

TALL WHITE BOG ORCHID

HELLEBORINE

LAKE HURON GREEN ORCHID

TALL NORTHERN GREEN
ORCHID

TUBERCLED ORCHID

HELLEBORINE
Epipactis helleborine   DOUG MCGRADY

31 Lip 2–3 lobed with no tubercle; statewide . . . .
. . . . **LONG-BRACTED ORCHID**
. . . . (*Dactylorhiza viridis*)

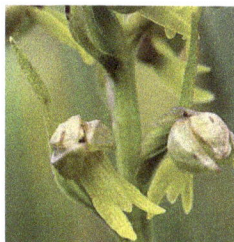

LONG-BRACTED ORCHID

32 Plants with one basal leaf. **33**
32 Plants with two or more basal leaves. **39**

33 Leaves linear **34**
33 Leaves wider, blunt-oval or round-oval **35**

34 Flowers solitary, magenta, leaf appearing after flower has matured; statewide . . . . **ARETHUSA**
. . . . (*Arethusa bulbosa*)
34 Flowers 3–12 in a loose raceme, magenta; statewide
. . . . **GRASS PINK**
. . . . (*Calopogon tuberosus*)

ARETHUSA

35 Leaves blunt-oval, flowers 3–15 in raceme, greenish-white; Upper Peninsula and northern Lower Peninsula . . . . . **SMALL NORTHERN BOG ORCHID**
. . . . (*Platanthera obtusata*)
35 Leaves broader, round-oval. **36**

GRASS PINK

36 Flowers solitary, rose-purple, white and yellow; leaf wrinkled; Upper Peninsula and northern Lower Peninsula . . . . **CALYPSO**
. . . . (*Calypso bulbosa*)
36 Flowers in a raceme. **37**

**SMALL NORTHERN
BOG ORCHID**

GRASS PINK
Calopogon tuberosus
JOSHUA MAYER

CALYPSO

37 Leaf present at flowering, not over-wintering; flowers 3–12, pinkish-mauve-white, lip 2-lobed; Upper Peninsula and northern Lower Peninsula . . . .
. . . . **SMALL ROUND-LEAVED ORCHID**
. . . . (*Galearis rotundifolia*)

37 Leaves green over winter; withered by time of flowering; flowers more numerous; greenish or yellowish, infused with purple; Lower Peninsula **38**

SMALL ROUND-LEAVED ORCHID

38 Leaf green, strongly pleated; flowers without nectar spurs Flowers 8–55, purplish-green-yellow, lip 3-lobed; Lower Peninsula . . . . **PUTTYROOT**
. . . . (*Aplectrum hyemale*)

38 Leaf green with purple spots on upper surface, purple on underside; nectar spurs present on flowers; rare in southwestern Lower Peninsula (Berrien County) . . . . **CRANEFLY ORCHID**
. . . . (*Tipularia discolor*)

PUTTYROOT

39 Plants with two basal leaves **40**
39 Plants with more than two basal leaves **46**

40 Leaves oval-lance shaped **41**
40 Leaves broader **42**

41 Flowers 5–27, racemed, mauve-green; southern Lower Peninsula . . . . **LARGE TWAYBLADE**
. . . . (*Liparis liliifolia*)

CRANEFLY ORCHID

41 Flowers 2–12, racemed, whitish or yellow-green; statewide . . . . **LOESEL'S TWAYBLADE**
. . . . (*Liparis loeselii*)

42 Leaves oval **43**
42 Leaves broadly oval to nearly round **44**

LARGE TWAYBLADE

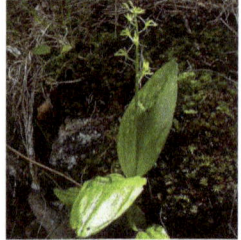

LOESEL'S TWAYBLADE

43 Flowers solitary, magenta, pouched; statewide ... . **PINK LADY'S-SLIPPER** .... (*Cypripedium acaule*)
43 Flowers racemed, pink-mauve with white lip; western Upper Peninsula and Lower Peninsula .... **SHOWY ORCHID** .... (*Galearis spectabilis*)

PINK LADY'S-SLIPPER

44 Leaves 3–5 inches in diameter, flowers numerous, greenish-yellow, lip and spur taper to a point; statewide................... **HOOKER'S ORCHID** .... (*Platanthera hookeri*)
44 Leaves larger, flowers numerous, whitish-green, lip oblong..................................... **45**

SHOWY ORCHID

45 Leaves 4–7 inches in diameter, spurs 16–27 mm long, flowers numerous, whitish-green, lip oblong; statewide ..... **LESSER ROUND-LEAVED ORCHID** .... (*Platanthera orbiculata*)
45 Leaves 5–8 inches in diameter, spurs 29–43 mm long, flowers numerous, whitish-green, lip oblong; northern Michigan .... **GREATER ROUND-LEAVED ORCHID** .... (*Platanthera macrophylla*)

HOOKER'S ORCHID

46 Leaves form a rosette, bluish-green, velvety, conspicuously marked with white, flowers white .. **47**
46 Leaves green, never marked with white, flowers white..................................... **49**

47 Flower spike one-sided **48**
47 Flower spike not one-sided................... **50**

LESSER ROUND-LEAVED ORCHID

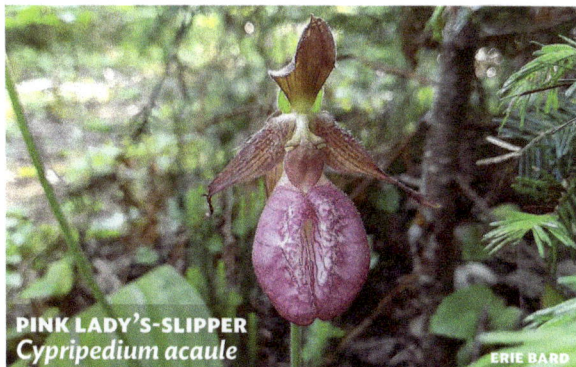

PINK LADY'S-SLIPPER
Cypripedium acaule
ERIE BARD

GREATER ROUND-LEAVED ORCHID

48 Plants 4–10 inches tall, leaves veined with network of white, lip tapers to sharp point; statewide . . . . **LESSER RATTLESNAKE PLANTAIN** . . . . (*Goodyera repens*)

48 Plants 14–18 inches tall, white markings on leaves faint or absent, lip tapering to blunt tip; northern Michigan. . . . . **GREEN-LEAF RATTLESNAKE PLANTAIN** . . . . (*Goodyera oblongifolia*)

LESSER RATTLESNAKE
PLANTAIN

49 Plant 5–10 inches tall, spike an all-round spiral, lip oblong, sac-like; statewide. . . . . **CHECKERED RATTLESNAKE PLANTAIN** . . . . (*Goodyera tesselata*)

49 Plant 6–16 inches tall, spike cylindrical, blunt-topped, lip rounded and inflated; statewide . . . . **DOWNY RATTLESNAKE PLANTAIN** . . . . (*Goodyera pubescens*)

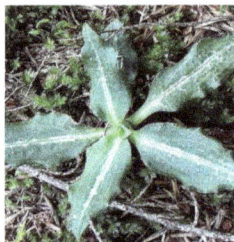

GREEN-LEAF RATTLESNAKE
PLANTAIN

50 Leaves persistent, not withering at flowering time, lance-oblong, smooth, fleshy, shining, spike crowded, several ranked, spirally-twisted; mostly Lower Peninsula . . . . **WIDE-LEAVED LADIES'-TRESSES** . . . . (*Spiranthes lucida*)

50 Leaves withering at or before flowering time . . . **51**

CHECKERED RATTLESNAKE
PLANTAIN

DOWNY RATTLESNAKE
PLANTAIN

MENZIES' RATTLESNAKE PLANTAIN
Goodyera oblongifolia

J BREW

WIDE-LEAVED LADIES'-
TRESSES

51 Leaves oval, spike long, single-ranked; statewide . . . . **SLENDER LADIES'-TRESSES** . . . . (*Spiranthes lacera*)

51 Leaves linear-lance shaped, spike 3-ranked; statewide . . . . **SPHINX LADIES'-TRESSES** . . . . (*Spiranthes incurva*)

SLENDER LADIES'-TRESSES

SPHINX LADIES'-TRESSES

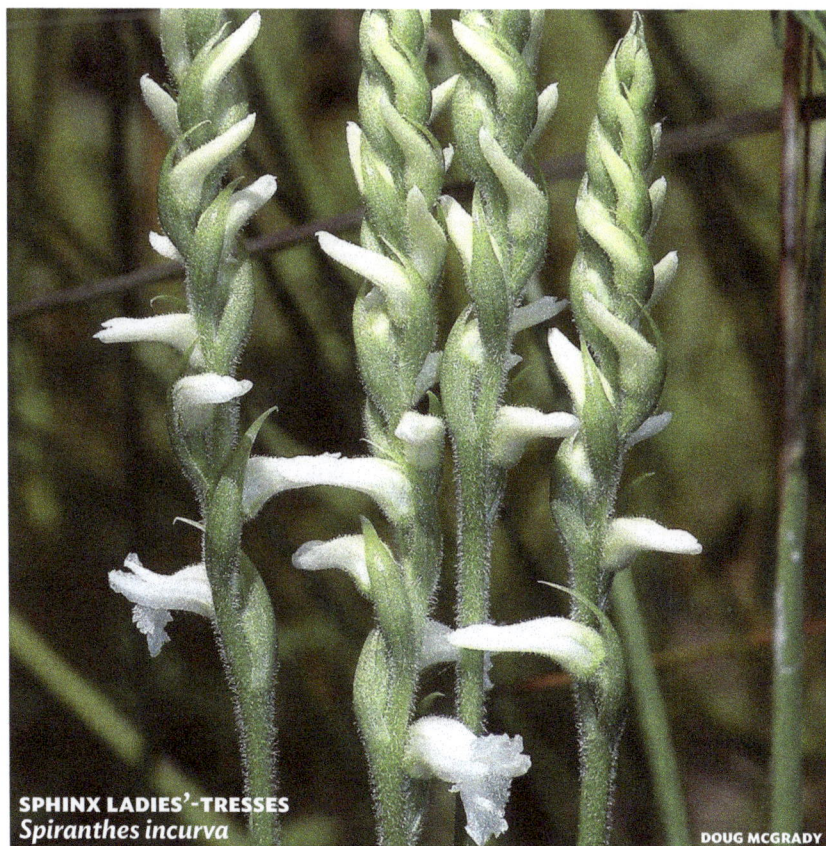

SPHINX LADIES'-TRESSES
Spiranthes incurva

DOUG MCGRADY

KEY TO MICHIGAN SPIRANTHES

NOTE Key adapted from Pace and Cameron (2017).

1   Lip strongly pandurate (shaped like the body of a fiddle); lateral sepals and dorsal petals appearing joined for most of their length, forming a 'hood'; statewide (especially northern Michigan)...... *Spiranthes romanzoffiana*
1   Lip not pandurate (but may be constricted); lateral sepals and dorsal petals separate, not appearing joined.. 2

2   Viscidium ovoid, column white; statewide (especially southern Michigan) .... *Spiranthes lucida*
2   Viscidium linear, column green....................................... 3

3   Lip centrally green or white with green veining; statewide (especially northern Michigan) .... *Spiranthes lacera*
3   Lip centrally yellow or entirely white 4

4   Flowers 4.5 mm long and smaller; southern Lower Peninsula............ .... *Spiranthes tuberosa*
4   Flowers 5 mm long and longer 5

5   Lip centrally papillate (covered with small bumps); mostly southern Lower Peninsula .... *Spiranthes magnicamporum*
5   Lip not centrally papillate... 6

6   Lateral sepals cupped... 7
6   Lateral sepals flattened.. 8

7   Lip centrally white; rare in southern Lower Peninsula ... *Spiranthes ovalis*
7   Lip centrally yellowish; mostly northern Michigan....... *Spiranthes casei*

8   Lip margin undulating; mostly southern Lower Peninsula .... *Spiranthes magnicamporum*
8   Lip margin crisped and lacerate 9

9   Lateral sepals downwardly sickle-shaped, their tips pointing toward the tip of the lip; reported for extreme southern Michigan . *Spiranthes arcisepala*
9   Lateral sepals sweeping upward, their tips pointing toward dorsal sepal and petals... 10

10   Upper surface of lip yellow, upper surface glands rounded; Lower Peninsula.. *Spiranthes ochroleuca*
10   Upper surface of lip white or very pale yellow, upper surface glands conical and reduced; common statewide.................... *Spiranthes incurva*

GLOSSARY

Anther, the part of the stamen which holds the pollen.

Bract, a modified leaf usually associated with the flower.

Calyx, the outer series of floral leaves.

Capsule, a dry fruit which splits open when ripe.

Cohesive, having a tendency to stick together.

Column, an organ (unique to the Orchid Family and several other plant families), formed by the union of the stamens and the style.

Corm, the enlarged solid bulb-like fleshy base of a stem.

Corolla, the petals of a flower, collectively; the inner series of floral leaves, usually colored.

Deflexed, bent abruptly downward.

Embryo, the undeveloped plant within the seed.

Endosperm, nutritive tissue within a seed.

Fungus, a plant of a lower order without stem, leaves, flowers, and chlorophyll, as mushrooms, mildews and molds.

Genus, a group of closely related species.

Herbarium, a collection of dried, pressed, and permanently preserved plant specimens, arranged systematically.

Lanceolate, shaped like a lance head, several times longer than wide, broadest above the base and narrowed to the apex.

Linear, long and narrow with parallel margins.

Labellum (or **lip**) the part of the flower that serves to attract insects, which pollinate the flower, and acts as a landing platform for them.

Nectary, a place where nectar is stored.

Ovary, that part of the pistil which contains the seeds.

Ovate, egg-shaped.

Papillae, minute conical hairs.

Petal, a division of the corolla.

Pistil, seed bearing organ of a flower.

Pollinia, coherent masses of pollen grains.

Proboscis, tubular prolongation of an insect's head through which nectar is sucked.

Pubescence, a covering of soft short hairs.

Raceme, a type of flower cluster as that of the Lily-of-the-Valley.

Rhizome, an underground stem sending leaves from the upper surface and roots from the lower.

Rootstock, a rhizome.

Rostellum, a little beak; a slender extension from the upper edge of the stigma.

Saprophyte, a plant which lives on dead or decaying organic matter.

Sepal, a division of the calyx.

Sessile, without a stem.

Species, a group of living individuals possessing common characters, the subdivision lower than genus in botanical classification.

Sphagnous, abounding in bog moss.

Sphagnum, a bog moss.

Stamen, one of the pollen-bearing organs of a flower.

Stigma, that part of the pistil which receives the pollen grains, usually the apex of the pistil.

Stolon, a runner that is inclined to root.

Style, that portion of the pistil which connects the stigma and the ovary.

Tubercle, a small knoblike prominence.

Tuberous, having the nature of a short, fleshy underground stein.

Viscidium, a sticky part of the rostellum that is removed with the pollinia as a unit and serves to attach the pollinia to the insect.

Whorl, an arrangement of leaves or other organs in a circle around a stem.

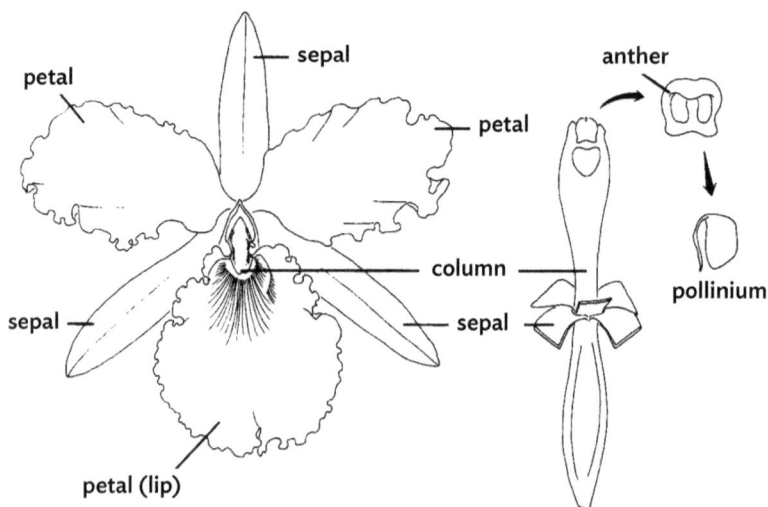

REFERENCES

Ackerman, J. D. & M. R. Mesler. 1979. Pollination biology of *Listera cordata* (Orchidaceae). American Journal of Botany 66:820-824.

Ames, Oakes. 1924. An enumeration of the orchids of the United States and Canada. Boston, 120 pp.

Barker, Elmer Eugene. 1931. You also can grow orchids as I do in New York State. Better Homes and Gardens, April, p. 13.

Boyden, T. C. 1982. The pollination biology of *Calypso bulbosa* var. *americana* (Orchidaceae): Initial deception of bumblebee visitors. Oecologia 55:178-184.

Brown, P. M. 2003. The Wild Orchids of North America, North of Mexico. University Press of Florida, Gainesville, FL.

Brown, P.M. 2006. Wild Orchids of the Canadian Maritimes and Northern Great Lakes Region. University Press of Florida, Gainesville, FL.

Case, F. W. 1964. A hybrid twayblade and its rarer parent, *Listera auriculata*, in northern Michigan. Michigan Botanist 3:67-70.

Case, F. W. 1983. *Platanthera* × *vossii*, a new natural hybrid orchid from northern lower Michigan. Michigan Botanist 22:141-144.

Case, F. W., Jr. 1987. Orchids of the western Great Lakes Region. Revised ed. Cranbrook Institute of Science, Bull. 48. 251pp.

Catling, P. M. 1976. On the geographical distribution, ecology and distinctive features of *Listera* × *veltmanii* Case. Rhodora 78:261-269.

Catling, P. M. 1980. Rain-assisted autogamy in *Liparis loeselii* (L.) L. C. Rich. (Orchidaceae). Bulletin of the Torrey Botanical Club 107:525-529.

Catling, P. M. 1983a. Autogamy in eastern Canadian Orchidaceae: a review of current knowledge and some new observations. Naturaliste Canadien 110:37-53.

Catling, P. M. 1983b. Pollination of northeastern North American *Spiranthes* (Orchidaceae). Canadian Journal of Botany 61:1080-1093.

Catling, P. M. 1983c. *Spiranthes ovalis* var. *erostellata* (Orchidaceae), a new autogamous variety from the eastern United States. Brittonia 35:120-125.

Catling, P. M. & J. E. Cruise. 1974. *Spiranthes casei*, a new species from northeastern North America. Rhodora 76:526-536.

Catling, P. M. & G. Knerer. 1980. Pollination of the small white lady's-slipper (*Cypripedium candidum*) in Lambton County, Southern Ontario. Canadian Field Naturalist 94:435-438.

Chapman, W. K. 1997. Orchids of the Northeast: A field guide. Syracuse University Press, Syracuse, NY.

Chestnut, V. K. 1898. Principal poisonous plants of the United States. U.S.D.A., Div. of Botany, Bull. 20, pp. 19-2o.

Correll, D. S. 1950. Native orchids of North America north of Mexico. Stanford University Press, Stanford, CA.

Darwin, Charles R. 1904. The various contrivances by which orchids are fertilized by insects. 2nd ed. 7th Impression. London, 300 pp.

Davis, R. W. 1986. The pollination biology of *Cypripedium acaule* (Orchidaceae). Rhodora 88:445-450.

Denslow, H. M. 1927. Native orchids in and near New York. Torreya, Vol. 27, No. 4, July-August, pp. 61-63.

Dieringer, G. 1982. The pollination ecology of *Orchis spectabilis* L. (Orchidaceae). Ohio Journal of Science 82:218-225.

Duckett, C. 1983. Pollination and seed production of the ragged fringed orchis, *Platanthera lacera* (Orchidaceae). Honor's thesis, Brown University, Providence, RI.

Flora of North America Editorial Committee. 2002. Flora of North America, North of Mexico. Volume 26: Magnoliaphyta: Liliidae: Liliales and Orchidales. Oxford University Press, New York. 723pp.

Fuller, Albert M. 1933. Studies on the flora of Wisconsin Pt. I: The orchids; Orchidaceae. Bull. Pub. Mus. of the city of Milwaukee, Vol. 14, No. I. Milwaukee, 284 pp.

Gleason, H. A., and A. Cronquist. 1991. Manual of Vascular Plants of Northeastern United States and Adjacent Canada. Second edition. The New York Botanical Garden, Bronx. 910pp.

Guignard, J. A. 1886. Insects and orchids. Annual Report of the Entomological Society of Ontario 16:39-48.

Guignard, J. A. 1887. Beginning an acquaintance with wild bees. Annual Report of the Entomological Society of Ontario 17:51-53.

Henkel, Alice. 1906. Wild medicinal plants of the United States. U.S.D.A., Bur. Plant. Ind., Bull. No. 89, pp. 23, 25, 52.

Holm, Theodore. 1904. The root-structure of North American terrestrial Orchideae. Am. Jour. Sci., Ser. 4, Vol. 18, No. 105.

Hogan, K. P. 1983. The pollination biology and breeding system of *Aplectrum hyemale* (Orchidaceae). Canadian Journal of Botany 61:1906-1910.

Holmgren, N. H. 1998. Illustrated Companion to Gleason and Cronquist's Manual. Illustrations of the vascular plants of Northeastern United States and adjacent Canada. New York Botanical Garden, Bronx. 937pp.

Homoya, M. A. 1993. Orchids of Indiana. Indiana University Press, Bloomington. 276pp.

Kallunki, J. A. 1976. Population studies in *Goodyera* (Orchidaceae) with emphasis on the hybrid origin of G. tesselata. Brittonia 28:53-75.

Kallunki, J. A. 1981. Reproductive biology of mixed-species populations of *Goodyera* (Orchidaceae) in Northern Michigan. Brittonia 33:137-55.

Keenan, P. E. 1992. A new form of *Triphora trianthophora* (Swartz) Ryd., and part 3 of observations on the ecology of *Triphora trianthophora* (Orchidaceae) in New Hampshire. Rhodora 94:38-42.

Keenan, P. E. 1998. Wild Orchids Across North America. Timber Press, Portland, OR.

Lownes, Albert E. 1926. *Triphora trianthophora*. Addisonia, Vol. 11, pp. 61-62.

Luer, C. A. 1975. The native orchids of the United States and Canada. New York Botanical Garden, New York, NY.

MacDougal, D. T. 1895. Poisonous influence of various species of *Cypripedium*. Bulletin of the Geological and Natural History Survey of Minnesota 9:450-451.

Morris, Frank and Eames, Edward A. 1929. Our wild orchids. New York, 464 pp.

Mosquin, T. 1970. The reproductive biology of *Calypso bulbosa* (Orchidaceae). Canadian Field-Naturalist 84:291-296.

Pace, Matthew C.; Cameron, Kenneth M. 2017. The Systematics of the *Spiranthes cernua* species complex (Orchidaceae): Untangling the Gordian Knot. Systematic Botany. 42 (4): 640–669.

Robertson, C. 1928. Flowers and insects. Carlinville, IL.

Rolfe, R. A. 1912. Evolution of the Orchidaceae. Orchid Rev., Vol. 20, No. 236, pp. 225-228.

Sheviak, C. J. 1974. An introduction to the ecology of the Illinois Orchidaceae. Illinois State Museum, Springfield, IL.

Sheviak, C. J. 1982. Biosystematic study of the *Spiranthes cernua* complex. New York State Museum Bulletin 448.

Sheviak, C. J. 1991. Morphological variation in the compliospecies *Spiranthes cernua* (L.) Rich.: ecologically-limited effects of gene flow. Lindleyana 6: 228–234.

Sheviak, C. J. and P. M. Brown. 2002. *Spiranthes*. Pp. 530–545 in Flora of North America vol. 26, eds. Flora of North America Editorial Committee. New York: Oxford University Press, U.S.A.

Sheviak, C. J. and M. L. Bowles. 1986. The prairie fringed orchids: a pollinator-isolated pair. Rhodora 88:267-290.

Steele, W. K. 1995. Growing *Cypripedium reginae* from seed. American Orchid Society Bulletin 64:382-391.

Stoutamire, W. P. 1967. The floral biology of the lady's-slippers. Michigan Botanist 6:159-175.

Stoutamire, W. P. 1968. Mosquito pollination of *Habenaria obtusata* (Orchidaceae). Michigan Botanist 7:203-212.

Stoutamire, W. P. 1971. Pollination in temperate American orchids. pp. 233-243 in M. J. G. Corrigan [ed.], Proc. 6th World Orchid Conference. Sydney. Australia. Sydney: Halstead Press.

Stoutamire, W. P. 1974. Relationships of the purple-fringed orchids Platanthera psycodes and P. grandiflora. Brittonia 26:42-58.

Smith, Welby. 2012. Native Orchids of Minnesota, Univ. Of Minnesota Press.

Swink, F. and G. Wilhelm. 1994. Plants of the Chicago Region, 4th ed. Indiana Academy of Science, Indianapolis. 921pp.

Thien, L. B. & B. G. Marcks. 1972. The floral biology of *Arethusa bulbosa*, *Calopogon tubersosus* and *Pogonia ophioglossoides*. Canadian Journal of Botany 50:2319-2325.

Thien, L. B. & F. Utech. 1970. The mode of pollination in *Habenaria obtusata* (Orchidaceae). American Journal of Botany 57(9): 1031-1035.

Voss, E. G. 1972. Michigan Flora. Part I. Gymnosperms and Monocots. Bulletin of the Cranbrook Institute of Science and University of Michigan Herbarium. 488pp.

Voss, E. G. & R. E. Riefner. 1983. A pyralid moth (Lepidoptera) as pollinator of Blunt-leaf Orchid. Great Lakes Entomologist 16(2): 57-60.

Watts, V. M. 1932. A clever deceiver. Nature, Vol. 19, March, p. 153.

White, Edward A. 1927. American orchid culture. New York, pp. 9, 35, 42-45, 65.

Williams, S. A. 1994. Observations on reproduction in *Triphora trianthophora* (Orchidaceae). Rhodora 96:30-43.

Marjorie Bingham and Dudley Blakely installing an exhibit, Cranbrook Institute of Science, 1945.

ABOUT THE AUTHORS

Marjorie Tellefsen Bingham (1895–1979), received her Masters Degree in Botany from the University of Cincinnati before coming to work as a botanist at the Cranbrook Institute of Science from 1933 to 1946. Between 1934 and 1941, Marjorie conducted a plant survey of Michigan which led to the publication of this work on Michigan's orchids in 1939 and "Flora of Oakland County, Michigan: A study of physiographic plant ecology" in 1945. At Cranbrook, she expanded the botanical collection through her fieldwork, and also taught physiology at Kingswood School. Among her accomplishments, she served as chairman of the Botany Section, Michigan Academy of Science, was first President of The Michigan Wildflower Association (est. 1941), and was instrumental in shepherding Michigan's wildflower protective law through the state senate. She left Cranbrook in 1946 to pursue a career in education on the east coast.

Steve Chadde (1955–) is a botanist and plant ecologist who developed a passion for the natural world at an early age. His grandparents, German immigrants to the United States, were avid gardeners, and provided him with early memories in their garden and orchard. Following his education at the University of Wyoming and Montana State University, he worked in the Rocky Mountain and midwest regions as an ecologist and botanist for the U.S. Forest Service, the Montana Natural Heritage Program, the Nature Conservancy, and various consulting firms. He has published numerous books on plants, including his state floras for Minnesota and Wisconsin, and for the Upper Peninsula of Michigan. He is pleased to make this landmark work on Michigan's orchids available to a new generation in a completely revised and updated, full-color edition.

ACKNOWLEDGMENTS
Photos were used under commercial-use Creative Commons licenses, and my sincere appreciation is given to the photographers who have made their outstanding images available. North American distribution maps were modified from those presented by Flora of North America North of Mexico (online, *beta.floranorthamerica.org*). Michigan maps were generated from data provided by the Biota of North America Program (*bonap.org*).

INDEX

NOTE Synonyms are listed in *italics*.